Bleak Joys

CARY WOLFE, Series Editor

53 *Bleak Joys: Aesthetics of Ecology and Impossibility*
Matthew Fuller and Olga Goriunova

52 *Variations on Media Thinking*
Siegfried Zielinski

51 *Aesthesis and Perceptronium: On the Entanglement of Sensation, Cognition, and Matter*
Alexander Wilson

50 *Anthropocene Poetics: Deep Time, Sacrifice Zones, and Extinction*
David Farrier

49 *Metaphysical Experiments: Physics and the Invention of the Universe*
Bjørn Ekeberg

48 *Dialogues on the Human Ape*
Laurent Dubreuil and Sue Savage-Rumbaugh

47 *Elements of a Philosophy of Technology: On the Evolutionary History of Culture*
Ernst Kapp

46 *Biology in the Grid: Graphic Design and the Envisioning of Life*
Phillip Thurtle

45 *Neurotechnology and the End of Finitude*
Michael Haworth

44 *Life: A Modern Invention*
Davide Tarizzo

43 *Bioaesthetics: Making Sense of Life in Science and the Arts*
Carsten Strathausen

42 *Creaturely Love: How Desire Makes Us More and Less Than Human*
Dominic Pettman

(continued on page 193)

Bleak Joys

Aesthetics of Ecology and Impossibility

MATTHEW FULLER and OLGA GORIUNOVA

posthumanities 53

UNIVERSITY OF MINNESOTA PRESS
MINNEAPOLIS · LONDON

An earlier version of chapter 1 was published as "Devastation," in *On General Ecology: The New Ecological Paradigm in the Neocybernetic Age*, ed. Erich Hoerl and James Burton (London: Bloomsbury, 2017); reprinted by permission of Bloomsbury Academic, an imprint of Bloomsbury Publishing Plc. An earlier version of chapter 4 was published as "Worse Luck," in *Revisiting Normativity with Deleuze*, ed. Rosi Braidotti and Patricia Pisters (London: Bloomsbury, 2012); reprinted by permission of Bloomsbury Academic, an imprint of Bloomsbury Publishing Plc.

Copyright © 2019 by Matthew Fuller and Olga Goriunova

All rights reserved. No part of this publication may be reproduced, stored in a retrieval system, or transmitted, in any form or by any means, electronic, mechanical, photocopying, recording, or otherwise, without the prior written permission of the publisher.

Published by the University of Minnesota Press
111 Third Avenue South, Suite 290
Minneapolis, MN 55401-2520
http://www.upress.umn.edu

The University of Minnesota is an equal-opportunity educator and employer.

Library of Congress Cataloging-in-Publication Data
Fuller, Matthew, author.| Goriunova, Olga, author.
Bleak joys : aesthetics of ecology and impossibility / Matthew Fuller and Olga Goriunova.
Description: Minneapolis : University of Minnesota Press, 2019. |
Series: Posthumanities ; 53 | Includes bibliographical references and index. |
Identifiers: LCCN 2018061114 (print) | ISBN 978-1-5179-0552-1 (hc) |
ISBN 978-1-5179-0553-8 (pb)
Subjects: LCSH: Environment (Aesthetics) | Philosophy of nature.
Classification: LCC BH301.E58 F85 2019 (print) | DDC 111/.85—dc23
LC record available at https://lccn.loc.gov/2018061114

Dedicated to
Varvara

Contents

Acknowledgments / ix

Introduction / xi

Devastation / 1

Anguish / 25

Irresolvability / 51

Luck / 75

Plant / 93

Home / 121

Coda / 155

Notes / 161

Index / 185

Acknowledgments

Many people and places have provided contexts in which we have been able to develop this work, and we would especially like to thank Alexandra Anikina, Cecilia Åsberg, Rosi Braidotti, James Burton, Olga Cmielewska, Geoff Cox, Jennifer Gabrys, Annie Goh, Erich Hoerl, Patricia Pisters, Margerita Radomska, Bev Skeggs, and Zooetics—Nomeda and Gediminas Urbonas. Doug Armato, Gabriel Levin, and Cary Wolfe have each been immensely supportive and made it a pleasure to work with the University of Minnesota Press and the Posthumanities series.

Introduction

Bleak Joys is a book about ecological aesthetics. It is also a book about bad things. Ecological aesthetics attempts to develop an understanding of complex entities and processes, from plant roots, to forests, to ecological damage as dynamic processes of composition. As such, its approach to the aesthetic is an expansive one that is both hungrily sensual and abstract. As a book about bad things, it discusses conditions such as anguish and devastation, which relate to the ecological but are also constitutive of politics, the ethical, and the formation of subjectivities and beings. These combine in the present day at multiple scales and in many ways, but they are also too often avoided, considered finite or absolute, rendered indifferent yet totalizing, because we do not have the language to speak about them. *Bleak Joys* attempts to capture some of the modes of crisis that constitute our present ecological and cultural condition, and to reckon with the means by which they are not simply aesthetically known but aesthetically manifest.

This approach to ecological aesthetics is combined with an engagement with number, prediction, and the modes of knowledge that are assembled to reckon with complex formations that are beyond simple calculation. As such, the book combines what it draws from ecological aesthetics with a set of accounts of calculative power and the wider ontological conditions brought about by the technological encounter with supercomplex systems.

In order to do so, it must site itself at points of paradox, of ambivalence and the overlappings of multiple absences. The question of the formation of such points is one that haunts contemporary theory in its proliferation of terms, such as complicity, irony, embeddedness, situatedness, the folds of the baroque, economic or topological subsumption—as these are figured in variously social, economic, epistemic, cognitive, aesthetic, and political forms. *Bleak Joys* inhabits such a condition—not in the mode of lament but as a part of ontological structure of its argument, which draws on the work of Baruch Spinoza, Friedrich Nietzsche, Gilles Deleuze, and Rosi Braidotti, among others, yet asks along with this tradition whether affirmation is enough, and what is to be affirmed. Setting up an interplay between an ecological materialism that is necessarily bleak, mineral, and appreciative of disaster on the one hand and the inheritance of the monist theorists of affirmation that find potentials and actualizations of a joyful *conatus* in being on the other is part of the pattern that sustains this inquiry. As such a pattern, it does not therefore find itself in some implied median between the better and the worse but instead is a means of sprouting axes that cut into and compose realities.

This bleak joy is a way of thinking things that are commonly and culturally figured as negative without losing the force of their impact but also without succumbing to the luster of mere doom. Its relation to philosophies of affirmation is therefore complex. We draw on the formation of the ethico-aesthetic, via Mikhail Bakhtin and Félix Guattari, a conjugation of aesthetics and questions of powers and the interarticulations of beings, ecologies, and forms of life, in which sensation and perception, following moves toward embodied and ecological cognition, circulate and are active in wide ecological dynamics yet crystallized and instantiated at numerous scales of analysis, experience, and force. More broadly than resting solely within the domain of sensation and perception, the ethico-aesthetic relates to a question of polyphonic composition discussed more fully below. These formations are rarely arranged around the question of the aesthetic dimensions of "bad things," such as ecological damage or the sense of anguish. In order to fully understand the present—a time that, like some others, distinguishes itself partly in the invention of novel forms of destruction—this is a necessary task. Ecological aesthetics is thus not a mourning for an irreparable and singular loss—nor simply musing, in more or less verbose terms, on the unknowability and inexpressibility of it all—but a process of finding the cultural and scientific coordinates of the damned vivacity of the cosmos in terms that are more fully adequate to it.

Here then we draw on and contribute to the posthumanities' relation to the question of science, proposing approaches that synthesize scientific and cultural modes of reflection. *Bleak Joys* takes as a given that scientific, cultural, and philosophical approaches and inquiries are neither symmetrical nor always in every case of equal value but are in a mobile state of disequilibrium that renders them in a state of great excitatory capacity. The book draws on scientific work in ecology, plant sciences, systems theory, computing, and cybernetics in ways that are not merely illustrative of but foundational to our understanding of ecological aesthetics and of the condition in which the posthumanities are being forged.

But to return to the question: what then are these bad things? And how do they gnaw into or dispel other processes? This is something we try to work through in reference to particular cases or kinds: namely, in the chapters devoted to devastation, anguish, irresolvability, and home. These chapters draw on a set of theoretical resources to ask questions of the ad infinitum cascades of relations and entities that make the world. One way of framing these questions is indebted to the vocabulary of Spinoza: what corrodes powers of action? How do capacities and processes of becoming generate interference patterns that induce absences or sad passions that are also intensifiers of experience or perception? Further, how can such dynamics be thought in relation to patterns of becoming in what are understood as open systems rather than the perfect geometrically closed systems that characterized the advanced science of Spinoza's seventeenth century? Equally, once we attend to entities and processes that are not entirely those described by the forces of reason and passion—that is, that are not the usual subject, the human of a particular ilk, but perhaps a swath of forest, a patch of sea, a species, and more even than these, formations of things that do not qualify as subjects or objects—what are the consequences for ecological aesthetics?

Such questions are asked in the chapters that follow, for which we now provide a brief map for navigation.

DEVASTATION

Devastation is an attempt to describe the process of becoming of depletions, pollution, radiation, the emergence of collapses, and the metastases of ecological disaster. Devastation acts on the Deleuzian figure of the virtual to attenuate the plenitude of possible becomings. This is something different from simple death, the state of full catastrophe or negation. Neither does it correspond to what Sigmund Freud called the death drive. Devastation takes part in the redistribution of potentiality, in the shaping of differentiation not to necessarily annul things but to effect change that may continue being multiscalar, if drastic.

Lars von Trier's film *Melancholia* depicts a planetary-scale devastation, a mere collision, an alignment of trajectories that makes a couple of two planets; there is an incidental nature to devastations. But they are also woven in with other factors. Oil spills, such as those of the Deepwater Horizon platform or those spills that are repeated well past the point of obscenity in the Niger Delta, become part of a political gambit, a brinksmanship of the worst, in which tap-dancing on the edge of the abyss becomes part of the negotiations around further extractions of value. The inability to recognize devastations as unfolding processes that do not necessarily arrive at obliteration but are extended, manipulated, and gambled with means that these edges are too easily slipped over.

Catherine Malabou's concept of destructive plasticity, drawn from her work on neurology, is useful here in that it provides a means for describing the moment when something snaps and moves into another, diminished state.[1] Occurring at certain scales, and at such moments, a new sort of commons may be produced in nature; an example of this kind is that of the commons of plastics in oceans and air, something we are all, unevenly, able to share in. This commons also implies the problem of the witness: from whose position or by what techniques is it possible to sense or to know a process of devastation underway? Here, the chapter reflects on the epistemological dimensions of ecological destruction but also on the way in which knowledge of such things has in turn provided grounds for military uses of ecological problems as a way to ratchet tensions and destabilize locales and polities.

Alongside larger-scale devastations of waste dumps and scorched earths, there are other scales, tied into ecologies and economies of food. Amid all these cases, a key concern of the idea of devastation is to find a language with which to talk about a form of becoming that is of an inverse kind to that usually described: a theoretical task for the present woeful age of species and habitat destruction.

ANGUISH

Starting from an initial scene of the indifferent, almost accidental killing of an elderly man by a young girl in the film *Three Stories* by Kira Muratova, this chapter develops an argument around the nature of anguish as a mode of experience. Proposing that anguish can be seen as an ethico-aesthetic sensibility forged within depletions of the world but also in conditions of menacing vitality, it suggests an ecological mode of experience in distributed scales of gestation. The chapter contemplates the way in which anguish is inhabited beyond the purview of the human subject and, more broadly, becomes a condition of coming into being in dissonant reconfigurations of the possibilities of the future.

This account of anguish draws upon the Russian notion of *toska* made renowned by Vladimir Nabokov as a painful yearning or rending of being, which runs from boredom or lovesickness to fatal existential despair.[2] This anguish can be a form of *conatus*, a matrix of condensation of individuations, which is traditionally causally explained, valorized, and imbued with meaning. To this end, we trace Nietzsche's critique of the ideology of suffering in an attempt to know the distress of our time. The affirmation or nonaffirmation of such a sense is something drawn from the films of Muratova and from Varlam Shalamov's stories of life in the Gulag, in that a recognition of anguish does not necessarily imply a justification of it in the grammar of pain, nor an attribution of any positivity to it.[3] There is an arbitrariness to such an experience; anguish does not necessarily arise from properly formulated political conflicts or clearly expressed ecological tragedies but forms in-between the jarring and shearing of scales, in encounters with alterations and diminutions of future becomings, in which living matter finds itself. Finding a means to properly understand such a sense in the search for adequate ethico-aesthetic vocabulary is central to this chapter. It concludes with an inquiry, in some way started by Eugene Thacker,[4] into the place of anguish in relation to the philosophy of immanence, as developed by a line of thinkers including Spinoza and Deleuze, with its bliss that may turn out to be bleak.

IRRESOLVABILITY

Irresolvable problems are those, to paraphrase the novelist Christa Wolf, to which none of the answers available are the right ones. Irresolvability is the structural incapacity to sort out a problem, to be in a state of inhabiting a problem that both consists of you and that is outside you or that is experienced as such by agents of a strategic thought that occupies, under duress, the condition and place of a thought that might resolve it. It is a means of establishing a certain kind of economy of deterrence, dysfunction, a generalized condition of sludginess. Irresolvability names the condition in which the structuring incapacity of action of strategic thought becomes—by means of related technologies, economic and organizational forms, and processes of subjectivation—a part of everyday infrastructures of feeling. Born in the game-theoretical exuberance of the Cold War, irresolvability names the rationalized technique, inaugurated at Hiroshima, of rendering a problem beyond reasonable choice. It thus establishes a connection between ecological obliteration and the prohibition of thought by means of reason. This state of impossible choice becomes foundational to the modern world. It marks a shift in the nature of choice into one different to the affirmative choice presented as key to the existential systems of Jean-Paul Sartre or of Søren Kierkegaard, which is unidirectional (if also perhaps bad). You must make a choice, but none of the choices you will make is the right one. Gregory Bateson presents us with the form of the schizogenic double bind to articulate this condition.[5] It is a formal condition that may be replicated across species but also one that works across scales.

This condition has its manifestations in art. Pop Art, partially catalyzed from the twin aesthetic condition of irresolvability and ostensive abundance, crystallized it in the 1960s; and its form in the articulation of the incoherent image of multiple, never fully intersecting surfaces comes to the fore contemporarily in the work of artists occupied with the conditions of digital images. Here, clusters and swarms of discrete iconographic and representational forms stagger and fragment across the image plane of the screen.

But this condition is also that of a generalized proliferation of irresolvable problems, a world of debts that come before persons (which, in certain territories, are built into the architecture of personhood, such as education) and the fractal proliferation of deterrence that dissolves futures and in which formalisms, derived from the techniques of deterrence, become structuring devices that saturate everyday life. In this regard, the chapter develops a reading of the work of Friedrich A. Hayek and his work on economic automata as part of a generalized and nonlinear infrastructure of irresolvability. The figure of individual choice has been fundamental to the economic and political transformations of the last half century. The systems by which choice is arrived at and encoded necessarily contain a means by which it is also deterred. An underrecognized achievement of the contemporarily reigning but faltering forms of political economy is the systemic novelty by which such deterrence may be arrived at naturally, in the iterative determinations of an informational ecology, by the nonlinear emergence of self-organizing voids. The impotences thus arranged lay the disavowed foundations for the return of rage characterizing crucial ruptures in the present.

Reflections on the work of the playwright Sarah Kane and that of Christa Wolf close this chapter in their descriptions of situations that "can take away your life but not give you death instead" or that, revising Medea, replace the Kierkegaardian aporia of death in the name of the father with other forms of logic.[6]

LUCK

Chance—and its elaborations as risk, fate, and luck—is a key figure here: what are the odds on life? Who and what will have to perish in contemporary society's ecological gambles, structured through innumerable decisions, calculations, assessments, neglect, fear, and delirium? What in turn are the ways in which such structurations enter into and prefigure chance? Luck, as well as fate and risk, are forms of hypotheses. But they are also a means of directing, explaining, and experiencing the differing *ontological loads*, the variable exposures and ability to act upon a distribution of chance that cultures, ecologies, and moments undergo. This chapter aims to suggest that the actualizations of chance in figures of risk, fate, and luck are ethico-aesthetic figurations that preform, distribute, and stage operations of chance in the contemporary gambit of living.

The chapter first offers an account of the development of the figure of chance as an openness to the cosmos in Deleuze and in Jean Baudrillard's critique of Deleuze's "gambit of innocence" to be found in *Seduction*.[7] The simply anthropocentric character of the aesthetic of chance proposed by Baudrillard makes it less useful for the encounters with nature to be found in Charles Darwin and Nietzsche. We suggest a move away from descriptions of chance as something simply and always ontologically open but rather toward means of assaying it as something worked and textured by figurations and mechanizations of probability: often under the guises of prediction, risk, and entrainment that may be incidental as much as intended.

The chapter examines the way in which chance is systematized through artifacts, such as the Galton Board or more abstractly derived mechanisms, or in runs of consistency in economies and in games. In such conditions chance may be interpreted or structured as risk. Alongside this framing of the arbitrage of chance is fate, an ancient form of chance, often reserved for those rendered unlucky (often categorized as nonhuman or less-than-human), whom "fate" reserves to be trapped in ecological devastations and other larger- or smaller-scale disasters. Embracing fate means having to bite the loaded and splintering dice of chance in ecology in conditions of inequality. Luck, in turn, is employed as a descriptor of good fortune for those managing to refigure their futures among the various conditions of depletion. Luck's ambivalent capacities might coincide with other vectors and generate tamings of chance.

These activations and articulations of chance are a form of coercion as well as invention, of staging encounters with chance that are arranged and shaped variously by, among other things, species, ideas, and events. Ways of making home in bad luck and developing an art of inhabiting and structuring contingencies operating at different scales of power is a roll of the dice that closes the chapter.

PLANT

This chapter presents an attempt to work on the scientific accounts of plant intelligence and neurophysiology from scientists such as Anthony Trewavas and Frantisek Baluška in terms of ecological aesthetics.[8] These accounts address the degree to which a plant is able to respond to its environment, to mitigate or respond to problems. Following the particular example of roots navigating subsurface space in relation to immediate and general factors such as the presence of nutrients, soil texture, and gravity, it examines the internal mechanisms of the root to explore the botanical propositions for minimal intelligent behavior and suggest that these can also be reframed as aesthetic. We argue that, as a term for describing interactions between organism and environment, the aesthetic allows for an understanding of the sensual and perceptual, without implying a possible detour through complications associated with other terms such as "consciousness."

Bakhtin's formulation of the aesthetic as a means of recognizing and inhabiting the interplay of forces, including those that are unknown, is suggestive here and readily scales to ecological considerations. The chapter develops accounts of how the particular capacities of different species, and individual plants, establish dispositions to the environment that can be read through the work of scientists such as Barbara McClintock, Francis Hallé, and others but also in work by artist Mel Chin, whose *Revival Field* works with plants with high capacities for the uptake of heavy metals and other pollutants in landscape remediation. Such artworks are significant in establishing plants as aesthetic agents at multiple scales substantially beyond the simply visual.

After exploring a number of cases in botany, literature, and art, we propose two heuristics for thinking about vegetal aesthetics—fatalism and glory. Fatalism is a mode of embracing the interplay of determining forces. Glory is the capacity for organismic exuberance in the blooming, breeding, and voluptuous or intransigent growth of plants. One interplay between these two is to be found in the arabesque—a coiled repetition and scrolling pattern that entails the extension of the body of the plant to its furthest reach. It operates in variant forms across mathematics, art, and dance, and in familiar plants, such as ivy, wisteria, or honeysuckle, where the cells on the side of a tendril stimulated by touch slow down their growth in order to wrap around a support, as the nontouching side expands more rapidly.

We argue that an engagement with the aesthetics of plants developed in botany and art allows for a more substantial ecological configuration of cultural theory, and that were such aesthetics to be foregrounded as an aspect of ecological life, other kinds of nuanced analyses of plant behavior might be made.

HOME

The presiding figure of interest in this chapter is that of a home, a house by the forest, that allows for another kind of living. It appears as a figure in nearly all films by Andrei Tarkovsky but also in many other works of Russian cinema. This figure allows for work on the reformulation of the question of nature to be continued, a thread to be woven here through the rethinking of the familiar trope of "our home on planet Earth."[9]

This chapter starts and concludes by drawing and reflecting on aspects of the work of Vladimir Bibikhin, a Russian philosopher whose lectures are sparsely translated into English. One of his most interesting works, from the point of view of ecology, is called "Forest," and it is in relation to its formulations that the figure of home is initially developed.[10] Bibikhin offers a reading of forest as matter that expands from molecular to historic, from biological to psychological, from political to cosmic, in a form of attunement that pays attention to every scale, is inclusive of conflict, and is cognizant of dreariness. To make a home in the forest of matter, to live in relation to material capacities, involves a lot of work.

Such work in this chapter starts from unpicking various legacies of "home." In Bibikhin's reading of Georg Wilhelm Friedrich Hegel, looking for one's "own" home in genus, nation, and will exemplifies a tradition of vertiginous circular seeking for an ownness of being that always finds a hole in its heart. In this it has an affinity to the work of Martin Heidegger, who annexes being to the apparatus of the nation by using death as his axiological pivot. This idealist edge of the home of hearth, cradle, and country is offset by Deleuze and Guattari's cosmic figuration of the natal. These legacies make only one axis of thinking home.

Other axes include the political–poetic, in the work of Hannah Arendt and Gaston Bachelard, and reflections on these through migrant homes, explored in some formulations of Homi K. Bhabha and in work by writers Ágota Kristóf and W. G. Sebald. We discuss nativist, maternal existence in homes before moving on to an economic–ecological axis that explores the ecologies of forests in relation to histories of property, cultivation, freedom, scarcity, and the trade in fur.

Here, the ecological and ethico-aesthetic location of the image of a home in the forest is significant. The history of the Russian forest is connected to the limits of the regime of serfdom, which the territories of the forest to the north and east of the region of fertile land never endured. This is not a phenomenon to be romanticized but is part of the legacy that the home in the forest carries: here we have the establishment of the idea of forest as both a free space and one of exile, one that contains sources for survival but not labor, as in grueling agrarian work under proprietorship.

Such a space of freedom is, however, historically linked to the trade in sable pelts that contributed to state formation and expansion, and whose limit was species extinction.

The boundary conditions of different possibilities of plant growth and animal life connect the ethico-aesthetic foraging of plants and fauna to the political conditions of space. This chapter discusses tight intertwinements between economics and ecology, nationalisms and metaphysics, and in so doing offers a way of imagining making a home fitting current times, as a mode of thinking, doing, and living. Linking inside to outside, earth to cosmos, politics to poetics, the ground to the underground to the aerial, animals and plants to law and violence, ecology to economy, pasts to futures: our figure of home in the forest of matter is an ethico-aesthetic proposition, without holism, with which to conclude.

ACROSS CONJUNCTURES

Aside from each chapter's ostensive topic, a hope with this book is to develop some sense of conditions and ideas about them that move across and are adaptive amid other formations: some more or less explicit propositions of relations of entities that traverse each episode. We would hesitate to call such things a method, since many of them arise after the fact, the facts of thinking in writing and returning to what has accreted. Such propositions may run like streaks of a colorizing mineral in a slither of marble, integral, absolutely, to the substance, picking out in correlation clusters of other formations but slightly unexpected at times, and of an irregular distribution, the terms of which we are in turn trying to get at. Part of this marbling is a sense of the common awkwardness of the themes addressed. The scenes, conjunctions, and prefigurations these chapters explore are frequently things that are suspended in indecision, unseen, or written off. Alternately, they are often unfortunate circumstances that may be incidental or in the process of being mined for bitter value via instruments that inhabit, elicit, and entrain the kinds of problems and conditions toward which this book addresses itself. Part of the proposition here is that not only working with problems but inhabiting them is also sometimes part of the necessary work of establishing selection criteria against which they will be transformed. As such, the approaches, figures of thought, vocabulary, and operations we trace and try to experiment with change scales and consistency and travel across chapters, connecting them by burrowings, echoes, and translations between pages. A few of these are mapped below.

Both abstract dynamics and very material textures and conjunctures appear across chapters, and this is in part because naming them requires looking at their various facets. There is, for instance, something in common in contemporary ecological politics, and political economy more broadly, between devastation and irresolvability. The chapter "Devastation" deals with patterns of life unfolding in times of crisis, and as forms of crisis, that are often silent, slow, and partial. The discussion of devastation starts an exploration into a wider modus operandi of power, one that has a related texture, a related modality to that which is figured as a kind of external structuration as well as a form of subjectivation in "Irresolvability." This chapter looks at familial resemblances between operations that structure a range of conditions, from the doctrine of deterrence in nuclear warfare to climate damage, as irresolvable problems—and there is some common art to their construction. This condition can be partly expressed as the complex inhibition of the means to resolve problems but also as an attempt to name the characteristic becoming of ruinations.

The ecological aesthetic resources and sensibilities needed to live through devastations include "Anguish." The chapter devoted to this condition focuses on aesthetic experience forged amid negative productivity and develops, together with but differently to the "Plant" chapter, an ethico-aesthetic vocabulary that aims at working with figurations of the ecological tuned into formations of the future.

The figures of risk, fate, and luck in the chapter "Luck" are responses to the conditions and experiences explored previously. These are descriptions of causality and the active structurations of chance, which act as variant forms of constraint, competition, and impossibility, made in order to structure and explain, with greater or lesser degrees of determination, phenomena discussed across chapters. Here, inventive forms of technical, mythical, and mathematical action texture events and processes at scales that may be discreetly described as biological, epistemic, or cultural in ways that do not merely laminate the one onto the other but in which there is an interweaving of forms of composition that may be anything but melodic.

"Plant" links back to "Devastation" and "Anguish" in proposing a discussion of vegetal life, in the midst of contested and damaged ecologies, in terms of aesthetics. This concern echoes aspects of the approach adopted in "Irresolvability" that traces various kinds of compositional consistency across scales and also, together with "Home," offers a technique of understanding relations emerging in complex forms of organization, association, and commensality. "Home" continues exploring kinds of thought, politics, and being by talking about living beside the forest as a means of living in relation to material capacities more broadly. Questions about the possible ways of knowing, experiencing, feeling, and witnessing across and amid proliferating scales are explored across all chapters as they become the questions of the art of living at home on planet Earth.

Throughout, we adopt the term "climate damage" rather than "climate change" because the latter, posed in the form of a diplomatic indeterminacy, has two problems. First, it is too generous, letting the terms of the genesis of climate damage off the hook, setting it up as something merely to be contemplated as a simple change rather than an ongoing event of real urgency. Second, the term "climate damage" (like the word "pollution") has the capacity to name a phenomenon that is both a thing and an act. It is in both these terms, and thus in relation to its genesis in capitalism, that it must be halted and reversed.

ETHICO-AESTHETICS

One might imagine that there is a movement from the bad to the positive, that the lines of the book drift gently toward a healing power. Such a wholesome glide is not within its power. There will be no Hollywood version of this book with attractive stars finding redemption through abjection and difficulty. Instead, we aim to explore how *Bleak Joys* articulates something of a cultural ecology and an ethics of the tricky art of living. One of the threads for such an art is a development of *metis*, a form of wise cunning, named after the Greek goddess who exemplified this quality. This conjunction of wisdom with cunning implies interaction between a sensibility or an episteme and an aesthetics and power—but power as arrayed in the midst of things. It is for this reason that this book often returns to the term "ethico-aesthetics." This is an ugly compound word but one that seems especially useful despite this. Indeed, perhaps its ugliness forces one to think.

What does it mean to think an ethico-aesthetics in the *present* moment? Guattari's book *Chaosmosis* suggests, in what is referred to in its subtitle as "an ethico-aesthetic paradigm," the preeminence of a broad-ranging conception of aesthetics for understanding the current conjunction in culture, philosophy, politics, and life.[11] *Chaosmosis*, and Guattari's work in general—including its connections to relatively submerged currents, such as the work of Mikhail Bakhtin and later cybernetics in its "epistemological" phase—suggests that aesthetics becomes a crucial compositional force in the contemporary world.[12] But further, in its conjunctive form with ethics, it provides a means of slipping a few tumbrels on the polymorphous lock of understanding of the kinds of forces and conditions that are operative today. Ethico-aesthetics thus provides a means of recognizing the multifarious dynamics that must be taken into account and that are to be experimented with in the formation of politics and aesthetics as, in their mutant forms, they are currently found in the world.

In this regard, Deleuze and Guattari write in the affirmative. They do so to write themselves out of numerous orthodoxies, to create a space inside the shuttered grimness of the decade following that of the publication of *A Thousand Plateaus* and to recognize a recursive and mutable ontology of being that is constituted by difference, multiplicity, and the inevitability of the new. To build an ethico-aesthetics means also to work in relation to other conditions of such ontology. Here, ethico-aesthetics is deeply linked to the question of *physis,* of nature, and of ecology, and needs to be thought through at multiple scales of immanence, including those of fundamental forces such as chance or heat, in terms of potential disaster, as well as those of the intimate, public, intellectual, habitual, political, and aesthetic.

The notion of the ethico-aesthetic in Guattari works with an understanding of aesthetics that is prior to the separation of aesthetics from "life." That is to say that although it traverses fields, such as art, in which aesthetics is explicitly refined and worked on in relatively laboratory-like conditions, its domain has no a priori institutional or conceptual limit. This is of crucial interest because such an approach does not find itself wincing at the anticipation of capture or recuperation, or conversely relishing a saving purity, in the context of art. Rather, it recognizes such factors as part of a wider set of compositional dynamics, which are to be navigated and manipulated, ignored or indeed gambled with or endured.

One of the ways in which ethico-aesthetics is of value is the way it provides for a means of thinking about events, things, processes, and dynamics as they occur across more than one scale. The term "scale" is useful, because it allows for the analysis and figuration of dynamics that are self-consistent but that have more than one mode or context of manifestation. Each of these requires the recognition of a certain consistency. Ethico-aesthetics, in turn, is a transversal scale that provides a means for thinking about scales in an appropriately polyphonic manner, and one that is required by its own consistency to be alert to variation.

In a section of *Cartographies Schizoanalytiques* called "les ordonnées (ordinates) ethico-esthetiques," Guattari talks about Bakhtin's work in the essay "Aesthetics and Theory of the Novel," where (echoing Immanuel Kant) he describes three scales (ordinates) in literature, those of the enunciation, the cognitive, and ethical and the aesthetic conjoined.[13] The work of Bakhtin is important to that of Guattari because again and again it provides a model of thinking through lived interrelations between multiple scales. In *Chaosmosis,* Guattari reintroduces Bakhtin's list of the five dimensions in which poetry operates. The fifth on the list recapitulates the whole: "The feeling of verbal activity in the generative action of a signifying sound, including motor elements of articulation, gesture, mime: the feeling of a movement in which the whole organism together with the activity and soul of the word are swept along in their concrete unity."[14] For Guattari, this conception of poetry powerfully amasses corporeal and incorporeal universes of signification.

In Bakhtin's *Art and Answerability,* the pertinent scales are the "three domains of human culture": "science, art and life." As is well known, Bakhtin's interest in the work of art or text was in drawing together numerous elements of life, and indeed, the forms of knowledge pertaining to them (for instance, linguistics, philology, and the human sciences) in a conjoint mode that is also understood via relation to philosophy.[15] A brilliance of Bakhtin is in how he hones in on and amplifies the proliferation of things into understandings, interpretations, and the changes, intentions, slips, customs, and feints that arise in the becoming of an enunciation or other kind of expression.

The ethico-aesthetic, as in *Art and Answerability,* is that which unites and runs through these different ordinates or scales. Their relation to other scales becomes in turn fractal, bearing repeated characteristics of composition, of repetition and heterogenesis (and here we can say that Guattari benefits from advances in science that allow an understanding of recursive multiscalar patternings, which in his work move from the psychic to the social, to the aesthetic, to the systemic, to the ecological).

The way Bakhtin sets out a relation between ethics and aesthetics is peculiar to him, but it directly concerns the question of the composition of scales and the thresholds that arise between them. Usually, it takes place in intersubjectival terms as these are played out within a text such as a novel. The ethical and aesthetic event takes place as concentric circles set in motion, where one can be found inside the other; as a set of circles intersecting where the outside for one becomes an inside for the other under conditions of transformation; or as a Möbius strip where the one turns out to be the other and vice versa. The singular event of a life consists of myriad multiplicities of such acts. Both the ethical and aesthetic dimensions of an event require means of attending to them. Bakhtin describes the ethical aspect of such a condition quite forcefully: "You can't live by the completeness of yourself and the event, you cannot act; in order to live, you need to be incomplete, open for yourself—you need to be for yourself as yet to become, should not coincide with that you presently are."[16] In *Chaosmosis,* Guattari relates this attention to incompleteness to Jacques Lacan's *object petit a,* and in *Cartographies Schizoanalytiques* to the formulation of the partial object of Melanie Klein, an idea developed by D. W. Winnicott as the transitional object. These terms are useful in relation to the question of the way in which the incomplete implies a particular range of compositional valences and dynamics. In this book's chapter on home, there is a development of related ideas under the term "outwardness." But they are also perhaps more general, implying a turning to the outer (whether that be for—or in terms described as and that imply transformation into—nutrients, energy, sex, and information) as an essential aspect of the composition of ecologies.

Such an approach also has implications for the way in which writing and thinking might be done. Outwardness requires a different mode of writing from the inescapable partiality of amongness, discussed in the chapter on anguish. In Bakhtin's writing the mapping of the proliferation of implications and understanding is often accompanied by caveats, or definitions, in parentheses, as in, for instance, the early stages of *The Problem of the Text*. Each text is rephrased, spun by a modifying statement that amplifies, clarifies, or inflects it with a prismatic expansion, ramifying its implications as infused with heteroglossic tensions. And it is Bakhtin's search for means of articulating the specificity of every element of the world that draws Maurizio Lazzarato to describe him as a philosopher of the event. The event is conjured up in the moment of dialogic relation, a unique occurrence but also one made amenable, indeed possible (since Bakhtin primarily worked on literature), by the collective force of language. And it is here that Lazzarato sees Bakhtin as offering a new image of thought that is fundamentally ethico-aesthetic. Thought occurs in "an assemblage of evental relations between the body, the incorporeal, the brain and the other."[17] Polyphony is art's early grasping of this new understanding of thought as dialogic event among the goings-on of the world as event.

It is part of the argument of this book that such a figuration is fundamentally suited to linking an ethico-aesthetic recognition of polyphony directly to the question of the ecological. Here, a becoming occurs; it individuates, but it also produces effects in other individuations that cooccur. These are describable at multiple scales. For instance, if one is to figure an organism in polyphonic relation to its environment, it might be done in terms of an access to or absence of resources such as energy or food, a position in a carbon cycle, an evolution in terms of fitness to a changing habitat, or a capacity to be alluring to potential mates, among other means. Equally, the forms of becoming of impossibilities are rendered with acuity by such means. By impossibilities we mean phenomena such as emergent absences and blockages, the strategic and technical enablement of drudgery or nonoptimality, and an overabundance of depletion that characterize potentially determining aspects of the present.

Correspondingly, but not symmetrically, each scale implies a set of modes of knowledge that have developed to privilege certain forms of access to it. Modes of knowledge emerge, or can emerge, in tandem with scales of realities and have a dynamic relation to them. Such emergence requires work, much figuring out, and a certain sympathy in striving to become adequate to that which it attempts. Here it must be noted that scales, or their modes of interpretation, are by no means always hierarchically stacked and that there are shared abstract dynamics that move across and are composed by their constituents. In a certain sense, the aim of this book is to find means of adequately addressing a small set of these abstract dynamics and the conditions of their composition.

To return to the question of what it might mean to think ethico-aesthetics in the present moment—to make an account in text that has some kind of sensitivity to the world, that is able to translate its polyphony into some strings of words—is to attend to reality-forming effects that are often submerged, diffuse, or spoken of in other terms. One of the tendencies in this book then is to attempt to place some sensors in places that potentially discern what might be happening. In dialogue with other work in the posthumanities—which figures human bodies, ideas, and societies as seething with the life and capacities of myriad others—we aim to find this ecological aesthetics as also having some consequences for the particular conjuncture of thought and experience as it comes into combination with text, part of the medial condition of a book. It is a small and incomplete set of these consequences that are hopefully presented to you here.

Devastation

♦

What we refer to as devastation is not solely a kind of becoming of nothing in which the nothingness is produced by this or that becoming of some thing; neither are devastations simply a diminution of the stock of entities in the world or the finite number or range of things. Instead, devastation is a kind of becoming in which the virtual is attenuated, depleted in some way, drained of its capacity to be constituent. Devastation operates and couples with, protrudes from, and dissolves certain other kinds of becomings that are biochemical, military and economic, sociopolitical, technical and mediatic, among other things. Our framing of this condition is inclusive of the three domains of ecology referred to by Félix Guattari: mental, social, and environmental—ecologies beyond "nature"—but today devastation also takes on the overtones of events such as extinctions, their threatening immediacy and increasing intensity.[1] *Bleak Joys* faces the need to recognize and explicate anxious humans, the strategies of modern warfare, calculations of probabilities, a rainbow of waste molecules in water, carcinogens, plastic or high-fructose corn syrup–packed bellies, oil spills, the proliferation of dross disguised as information, among many other layers and registers. Devastations cut across these to produce something that exceeds their categorical limits.

It seems that in the discussions of extinction, taking place for instance in the accounts of deforestation or other destruction of natural habitats, the Aristotelian model of genus and the forking paths of classification adhered to in the Linnaean system still have significant traction on the public sense of the diminution of the variety of species, in turn endangering the ecological and social horizons of possibility. This is important, but something more is occurring. In conditions of devastation it is not a set of things becoming extinct under a category or idea that is thus itself transformed, effecting the others in a cascading logical fashion that uncannily follows a treelike formation, but it is something more substantial, an existing multiplicity, a differential, that fails to actualize, a potentiality that is wounded in a way that makes it implode, that makes it actualize a devastating becoming.

Deleuze draws upon the example of a lens described by Henri Bergson where the virtuality of all colors in white light are actualized to offer a range of blue, red, and green; one could ask what happens to color if the blue of the sky is no longer actualizable because the atmosphere has changed or disappeared.[2] What changes in the concrete universal of light that passes through the lens when there is no actualizable blue of the sky or of water?

BECOMINGS OF DEVASTATION

The philosophies of desire and of process wrote themselves out of the condition of subordination reinforced by the Hegelian tradition in terms of ideology, history, false consciousness, and the like by emphasizing becoming and difference rather than being and identity. One could say that they replaced a universe of "final perfection with static existence" (as Alfred North Whitehead abbreviates that of Descartes)—ontologically, a mechanical universe, in which the machine can fully come to a stop—with that of an ecology, of nonlinearity.[3]

In the preface to *Difference and Repetition,* Gilles Deleuze talks about the problem of rendering the argument for affirmation in relation to discussions of the negative, predicated upon more traditional philosophical tools such as doubt, criticism, opposition, and so on.[4] The figure of the beautiful soul, in this account, sees only the gorgeously ever-differentiating oneness of it all, and is unable to deduce a mode of living or a reading of politics. Drawing on this fissure, for the purposes of measuring philosophy on the scales of a form of politics, is a line of inquiry developed by Benjamin Noys in *The Persistence of the Negative.*[5] Our tack is different here in that we want to develop a discussion of how what is seen as negative, or inimical, may operate by means of dynamics that is often rendered as belonging more properly to an idea of the anthropically beneficent fluctuations of nature, in which guise vitalism may often come. Such a thing becomes an oxymoron: a lively devastating vitalism, the becoming of obliteration.

The conditions of the genesis of the actual are grounded in the virtual, a differential infinitely saturated with change, infinitesimal or infinitely large, multiple. The virtual is real but not yet, or ever, actualized. The virtual is also fully immanent and is necessarily affected by the actual, too;[6] otherwise the virtual would operate as an eternal transcendental idea, unattainable and unthinkable. In what follows we seek to create an ethico-aesthetic description for devastation as it manifests in the virtual-actual continuum in ways that in turn parch the virtual.

Devastation is a kind of ontological flexure on the process of actualization and change. Devastation may not necessarily diminish complexity, and its effect is larger than the calculation of possibility, on the planetary or cosmic scales. There is no point arguing whether devastation only affects the domain of Eukarya or Prokarya, thus covering all cellular life; it does not only necessarily affect the latter while leaving physical and chemical scales intact in their orderly described trajectories. A metaphysical devastation, a devastation of the virtual, arises from the piling-up of shearings of multiple scales. Actual devastation does not create the virtual by way of resemblance or analogy but necessarily feeds into it, producing it. Equally, devastations are not simply diminutions of things, or of the range of things, but are a kind of becoming in which the virtual is altered, diluted, or maybe enhanced in a sour way.

Devastation does not imply that there is an end to becoming or negation of affirmation but that there is a change to the becoming of the virtual. Devastation is becoming that seizes, eliminates, or radically changes the conditions of other becomings. The tendencies of devastation are not, however, necessarily anti-organismic or entropic or as such faithful solely to the axioms of thermodynamics. Devastation can generate novelty and complexity outside of diversity. Devastationally complex forms include the dynamic behaviors of new autoimmune diseases, complex harmful molecule compounds, cancerous growths, radiation, and accumulations of carbon dioxide, which do not eliminate complexity and wholeness in favor of randomness or a flat lack of differentiation but radically redistribute the shares of potentiality, shape planes of activity, and tangle with, impersonate, and swallow other processes of change. The active growth of devastation is not the individually unthinkable scope of the death of the individual or the overwhelming absences of pure nothingness; it is something to the side of such things, being devastatingly vital, active, and productive.

Devastation is sometimes akin to a geometrical progression; pollution links to surges in cancerous growth, sugar and fat to obesity, obesity to the increase in human mass, changes of pH to changes in jet streams—devastation refers to such complex couplings that in their scale and scaling capacities and their intensity have a grandiosity and systematicity to themselves. While devastation is a kind of change that affects the virtual and changes the processes of actualization with some systematicity, it can also be an abrupt and discontinuous cross-cutting change.

More than three decades later, the Chernobyl disaster is a relevant example: the sociopolitical effects engendered by radiation seem to have ensured that a lack of anthropogenic factors in the exclusion zone contributes to its relatively higher biodiversity with rare animals being spotted there. There is a window of instability in such radioactively charged biodiversity that allows certain other factors to prosper for a while amid other unfoldings: the biochemical effects of radiation interfere with the microbial and fungal ability to process biological decay, thus leading to the conservation of the dead.[7] As a result, thousands of trees lie undecayed in the same spot where they fell. This interference with the dead is of a different quality to that of work attributed to the afterlife: it is an arrest of death.

Jean-Hugues Barthélémy has written on Gilbert Simondon's formulation of "deadening" that "is contemporaneous with each vital operation as an operation of individuation."[8] He suggests that Simondon's positioning of death as a deposit can be altered to reflect an understanding of death as "the very heart of life," a position he finds confirmed by contemporary biology, where "cellular suicide plays an essential role in our body in the course of construction."[9] At the scale of cells, the death of certain of their member in a developing embryo is a precondition for growth as a process of the separation of bones, digits, and orifices. In a related way, Ray Brassier notes the way in which cellular specialization occurs in evolution, where a primitive organism "sacrifices" a part of itself to protect the rest from the external environment and to functionalize itself, thus making death an origin of life: the death that cannot be repeated in death itself.[10] Such a form of death—part of an endosymbiotic becoming, a link in a chain of becoming, or an excluded and unthinkable attraction core to being—is radically altered in devastation.[11] Devastationary death leads to something other than further life and the recouping of material resources into linked systems, the becoming of other states, or the pull of "originary" death. Devastationarily arrested deaths are multiplex, cutting across scales of interpretative frameworks or capacities of knowing.

At another scale, devastation as ecological event can be characterized as involving complex and manifold interactions across and within multiple kinds of entities and systems. Earth's history is marked by a number of massive, planetary-scale events. We know that there are ages on Earth when many things die, such as various glacial periods. In the Great Oxygenation Event, the evolution of photosynthesizing bacteria (believed to be oceanic cyanobacteria) generated a significant amount of free oxygen, obliterating many organisms and triggering the longest glaciation but also creating conditions for biological diversification by creating new energy resources, a new atmosphere, and the ozone layer, and so life in its familiar forms could evolve. Things (like free oxygen) have qualities that can be destructive for other things, and radiation is perfectly "natural" as part of matter. Devastation does not simply amount to the existence of destructive qualities themselves or destruction per se. Devastation relates to changing the conditions of becoming and can be a form of very active production, reconfiguring the relations between stability and change, expansion and contraction, wreaking havoc in chains linking habitats to cosmologies, such as those that move from the destruction of forest to the extinction of the languages of those who live in them, resulting in loss of ability to think in certain ways.

Above, we differentiate devastation from a pair of other conditions. Death in devastation is not the traditionally understood part of the pair of life and death, part of natural cycles, and patterns of growth and decay, nor is it the polar attractor of the death drive. Certainly, both of these conditions may take part in devastations. But devastations take things out of cyclical or determined states into proliferating conditions of depletion. In the case of Chernobyl, the afterlife and growth of radiation, to take another example, is the result of the disaster drastically depleted fungal and bacterial operations, resulting in part in the nonreturn of nutrients to the soil. Such change delinks the source of life in nonlife or other forms of life and alters the processes of becoming. This is not simply a deferral of a usual process with trees "stored" for later decay but an effect of radiation's arrest of death in life that is itself a kind of growth, a propulsive unfolding of things, for which we have no available ethico-aesthetic figures. One possibility for these trees is that they maintain this dry unrotted state, an expanse of excellent kindling, until the advent of a forest fire whose smoke and ashes would spread the radioactive material they store far beyond the current exclusion zone. This would be a growth, an affirmative becoming for radiation as a kind of devastation.

MELANCHOLIA OF OBLITERATION

In the discourse of natural history TV extravaganzas, as Donna Haraway puts it, "knowledge saves" via conservation, scientific understanding, and popularizations.[12] A mediation of survival is one means of ameliorating conditions of devastation. In the case of Lars von Trier's film *Melancholia*, however, there is nothing to be done.[13] A rogue planet is on a fatal and implacable collision course with Earth. One is obliterated, we are obliterated, they are obliterated—everyone and everything is obliterated, along with the planet. There is no possibility for reflection afterward and no prospective capacity to understand or sense obliteration.

Is *Melancholia* just a scary occurrence of the impossibility of thinking the Earth beyond human extinction or does it recount a differentiation in and from devastation: the differentiated becoming of the perishing of human species, animals, forests, flows, continents, the Earth as a totality of its destruction, or as a subset of planets as a whole? *Melancholia* obliterates Earth as a living planet, but it does not cancel out its physical matter, which is scattered in space and possibly left to drift as atomic rubble. Are we tempted as humans to simply bemoan the obliteration of the virtual that we equate with earthly human potentiality or is it indeed an imaginary act of thinking the perishing of the virtual of all matter, echoing in some way that of the ultimate fate of the universe and the ontology of energy embroiled in thermodynamics?

Obliteration thus sets out another margin from that of natural cycles within the bounds of which devastations become manifest. Obliteration brings us to the question of the void, finitude, the vastness of nothingness, and questions of cosmology, states, and conditions that we do not pursue here but use as a point of approximate measurement. Such conditions are, as writers such as Eugene Thacker explicate, rather tricky to formulate observations of.[14] At the same time, and as such, these conditions act as a rather convincing limit.

SPILLS

One of the most obvious and egregious of devastating abundances is that of oil spills, from grounded and broken tankers, faulty and unguarded valves on oil rigs, and the collateral damage implied by the development of new techniques such as fracking. By such means, the Earth, all surface, gets in touch with its inner self. How is it possible to enter into knowledge of such events?

The becoming of the Deepwater Horizon event, the momentous leak from a BP rig in the Gulf of Mexico in 2010, for instance, is interrogated by a range of mechanisms, including risk discourse as epitomized in insurance contracts and legal liability, the articulation of claims of environmental stewardship and the diminution of what the stakes of such might be, the technical language of oil-spill management and the attendant withering of the terms of the precautionary principle, and the media responses of the various companies involved, distinguishable by the variety of their more or less inept and mendacious quality.[15] All of these produce their own kind of grasp on and amplification of the event, even when they try to smother it. Indeed, perhaps what the urgency of a reckoning of devastation is partially driven by is how such conditions are supposedly resolved by such discourse, with a resolution holding it at bay, boxing it off, rather than attempting any more sustained understanding that might risk fundamental implications for oil as a commodity.[16]

Oil is tragic because at the same time as providing enormous power, it poisons those associated with it, however remotely. Indeed, part of the complexity of oil is its profound corruption of the discourses, persons, and institutions around it as they work around the impact of this immense energetic and toxic force. Such work includes the stabilization of certain forces (the capacity of getting energy) and the harnessing of others for certain kinds of gain or utilization at the same time as the marginalization of recognition of certain of its consequences (climate damage).

The tragic nature of oil is apparent in the frequent reports of the results of deliberate or accidental ruptures of oil pipelines in Nigeria and elsewhere, thus compounding the baleful consequences of large-scale gas flaring and generally haphazard and negligent treatment of ecological effects.[17] Spills are regular, obliterating the use of land for farming and as spaces of ecological succor. The abundance of such oil indulges a disregard even for its wastage, not to mention the differential withering and bloating effects on local life of the colonial powers of the oil companies.

When spills occur in the slums and shantytowns, people collect some of the oil in whatever containers are available. Frequently these spills result in conflagrations, killing and burning all those who had gathered to collect the oil in their meager containers. Each such event is a catastrophe, but the ongoing form of spills and the negligence with which they are operated render their qualities those of devastation in the way that their proliferation goes unchanneled. As devastation, such spills populate entire ecospheres. They change the capacity of parts of the surface of the earth to sustain life by smothering it in a substance from beneath its surface, one composed, of course, by organisms decayed under particular circumstances.

One of the significant contingent factors about the way in which devastations mesh with human societies is that their unfolding is frequently gamed, manipulated, or gambled on for political advantage. Devastations are political, and are drawn upon by meshings of rhetorical, calculative, juridical, economic, and sociopolitical forces and interests. This is something readily observable in the brinksmanship passing for advanced statecraft in the negotiations over climate damage. Perhaps because obliteration is unimaginable, unrepresentable, that which edges toward it is not yet it. Devastation becomes the negotiable continuum. The void is unimaginable; therefore, it acts as some solid, finite as a basis for nonnegotiation, as a state that we have not yet reached. Tap-dancing on the rim of an abyss that cannot be seen looks so convincing if the dancer herself cannot see the edge. What should be a convincing limit is seen as a foundation upon which what is imagined to be political and economic advantage can be made. A moral, if not conceptual or speculative, limit thus provides the grounds for speculation upon its transgression on the basis that gambles will be made on the idea that it cannot be transgressed.

DEVASTATION AS PERSONIFICATION

Discussing brain injuries and drastic neurological conditions, Catherine Malabou posits a "destructive plasticity" to describe physiological events in the brain that fundamentally change a person such as advanced dementia in Alzheimer's disease, severe strokes, split identities, and other phenomena. Malabou asks a double question founded in negativity, for which she aims to recoup the possibility, both in reason and in the capacity to recognize as fact, of pure negativity: "Is there a mode of possibility attached exclusively to negation? A possibility of a type that is irreducible to what appears to be the untransgressable law of possibility in general, namely, affirmation?"[18] Malabou recounts that in the history of thought, an ability to negate was always an affirmative gesture: something started by Kant as an ability to say "no" and described by Hegel as always having relied on the implication of saying "yes" to "no," and thus based on a rounding principle of doubling, or affirmation. Here, "categorical refusal is not possible." Negative possibility, which is formative for Malabou, is of another order: it "is neither affirmed nor lacking," "does not proceed either from rejecting or spitting out," "refuses the promise," "makes existence impossible," "prohibits . . . the other possibility." It is not coupled up with "another," with a future.[19] The structure that makes possible the trickery of affirmation, of this double negative that always pulls an affirmative out of its empty hat, is partially the effect of language in which a "no" always has a presence. In a certain sense this is a related problem to the unrepresentability, or the unknowability of the void. Destructive plasticity instead marks a break, a fundamental event around which there is no possibility for a flickering of meaning but instead a snuffing out of what had been a person's character. For Malabou, "Destructive plasticity deploys its work starting from the exhaustion of possibilities, when all virtuality has left long ago."[20] In this work, Malabou provides a significant means for the recognition of the devastations within the scale of a person, whose existence is made impossible, unfolding on many scales: within the brain, at that of memories, behavior, motor function, and so on. There is no a priori limit to the virtuality of a person other than its constituent coupling with actuality. What Malabou maps so well, although in different conceptual vocabulary to us, is the modalities of damage that may constitute such actuality at the scale of the brain.

DEVASTATION IN COMMON

At another scale, as Elinor Ostrom notes, devastations occur to commons and are not limited to any particular scale, size, or location.[21] Devastation in fact may sometimes be the only common we are left with, but perhaps these commons exist only as a disowned residuum, since everything else is to be owned. One of these is the waste commons of plastic. This is an enormous distributed and discontinuous entity closely allied to water, as it is its uncanny distribution in lakes, rivers, and oceans where large concentrations of plastics first became known.[22] Microplastics (particles less than 5 mm in diameter), plastic particulate waste, and microbeads (used in substances such as facial cleansers) have been found across the world, from tap water and mineral water to the rather more remote and disconnected Southern Ocean.[23] Of the Earth's eleven oceanic gyres, the five situated in the subtropical regions where winds and sea currents slacken draw in floating debris. Here, plastics are broken down by ultraviolet light and movement of the sea. Some may decompose into toxins such as polychlorinated biphenyl (PCB) and nonylphenols.[24] Entering into the bodies of fish, albatross, plankton, or other species, there is the generation of chains of consequences: first-order throttling and blockages as the plastic objects take up space in stomachs, making them unable to fully digest food once some plastic has lodged in their belly; and a set of second- and third-order poisonings, directly into the organism that ingests it, and into those that may eat its body.[25] Novel ecologies occur when such fragments of plastic, in fresh or salt water, provide new microhabitats for bacterial assemblages that are selected by their ability to cling to their typically smooth or degraded surfaces—with a depleted taxonomic range as a result.[26] With the entry of such relatively new kinds of entity, ecologies become unstable yet difficult to map due to the redistribution of life and nonlife. The question of what can be said to exist in this plenitude of cloggings is related to the question of the proof of an absence, of the new forms of death and life.

These are the inverse commons produced by the interactions of myriad sequences of cost-benefit analyses that externalize costs and responsibilities. Plastics are cheap to produce and in the immediate-use context are mainly hygienic. Who is responsible for placing it in the water? Aside from the distended imaginary of the oceans as an ever-replenishable dumping ground, there is an endless deferral of responsibility that acts as a self-organizing entity in itself: oil companies, plastic producers, product designers and manufacturers, retailers, consumers, states and regulators, recycling systems, water companies or organizations, sewage and filtration plants, testing regimes. The list goes on with a maddening recursivity into every decision, at which the possibility to do anything is deferred as an externality to another entity in the sequence: long-chain polymers combine with long-chain causalities to produce toxic commons. The creation of a commons of plastic at a physical level is thus mirrored and matched by an emergent political form: but this mirroring is convex; it pushes all capacity for its articulation to the edges, where—it is hoped—it asymptotically vanishes. Disavowal, apathy, and indifference have their own structural logics that are foundational to contemporary political formations.

In relation to a commons of devastation—such as that of the 1984 Bhopal gas leak, where those structurally least able to bear the burden of pollution have been gifted with the opportunity to freely have it absorbed by their flesh, water, ground, and children—new political subjects may arise in contestation of such conditions.[27] The position of the waste picker—in which laborers work rubbish dumps, tide lines, and disposal sites—requires objects that are immediately graspable as such: old computers, scrap metals, hulks. They can be dismantled and smelted for purification that moves the undesired parts into smoky air. When the waste object is leaked into and now constitutes one's self, other conditions prevail or overlap.[28] There is a certain affinity here with the way in which uncanny or alien forms may flourish in the zone, as described by Boris Strugatsky and Arkady Strugatsky in *Roadside Picnic*.[29] The devastated zone has another capacity of becoming, and its potency as a mutational field is what is most stunning in Andrei Tarkovsky's *Stalker*, the film based on *Roadside Picnic*.[30] Such a response to devastation is part of what art often offers, a material imagination of adaptation, mutation, or horror—an aesthetic parallel to evolutionary models of endosymbiosis, commensualism, and parasitism—that allows for these conditions to be sensed.

But the problem with human culture in relation to the manifestations of devastation is that it is so often stuck in the positive, the little twinkle of redemption at the end of every 35 mm apocalypse. There are very few aesthetic figures (film director Kira Muratova, whose work we explore in the next chapter, is an excellent counterexample) that can contemplate the dark without drawing a resolutely positive lesson, taking the time to watch or to stare without jerking back. Perhaps this is a lasting legacy of Judeo-Christianity, transformed into the Anglo-Saxon gift of compulsive optimism. Conceivably, in turn, the belief in the intensive and vital capacity of change sometimes has some wishful thinking to it, one that is less that of a conceptual and aesthetic imagination enacting and invoking new worlds than a soothing tale of things sorting themselves out in a jolly cosmos where irresolvability, futility, and meager meaning do not figure.

To phrase this concern more in terms of a question: how can difference be contaminated or ground to a halt by too much difference? How might a philosophy of difference account for plastic in albatrosses' bellies—the cross-cutting of systems that yield plastics and yield albatrosses? A few intersections of these things create devastating conditions, whose intensive character or networking of scales do not bring about a fruitful transition to another state but inflect actualizations that, while destroying the actual, also parch the virtual.

Devastation is not always an entirely catastrophic event. It can be slow, familiar, pleasant (sugar dumps in bodies), or unnoticeable; it can be cumulative, mutational, or depletive only for a second or third generation. The nonlinear causality of devastation holds but does not create complex things of wonder as the various machines of evolution or thermodynamic systems far from equilibrium are said to do. Devastation creates something for which we have no image.

WITNESSES AND WARFARE

Such a condition suggests something that deserves to be recognized. As it moves across scales, devastation requires a sliding subject—some forms of thought and mechanisms of observation that are able to follow scales, registers, atoms, organisms, habitats, languages, chemical compositions, pain, hunger, changes of structure, shifts, and torsions of power and possibility—in order to carry out such recognition. Thinking about such a subject, Malabou asks, What might be a phenomenology of damage?[31] In this is nested a clutch of questions. Within the scales of destructive plasticity and the richly varied susceptibility to damage of the brain, who or what and with what instrumentation and sensitivities is there to make an account of such an event? Since there is not always the other, who can make an account of the ways a devastating change becomes manifest? Not even a self is capable of marking the ways in which it became other to itself—what modes of witnessing are then adequate to devastation? Is devastation always happening to an entity to which such an undergoing can be delegated and deferred?

In materialist ontologies, suffering, diminution, pollution, cancerous growth, changing pH levels, melting ice, and the evaporation of lakes have scales and modes of existence larger than those that humans can conceive, experience, and project on their own. Just as there is to life, there is an incomprehensibility to devastation. A problem for ecological science today is trying to comprehend: from experience, from imagination and modeling, from a fastidious tasting of samples of core ice, tree rings, atmospheric gases, and climatic records—trying to understand the roots, conditions, and counterfactuals of the incomprehensibleness itself. The problem of who or what is thinking and watching the devastation and for whom, at which scale it occurs, means also trying to establish the means by which such accounts can be elicited, at the same time as recognizing that a full unfolding of the condition is unknowable.[32] The simultaneously empirical and abstract status of devastation is a problem. It is one that calls for an abstract empiricism, one capable of making a reflection on the constitution of such problems on multiple levels and scales. Perhaps it is one that might resonate with contemporary physics' figurations of the multiverse,[33] in which myriad coexisting universes all require their own well-equipped observers or poets, only a few of which happen to be humans. But this is not simply a problem for thought and its iteration on a solely philosophical or scientific-technical level. As devastations may not be so *evidently* extreme and are sometimes not quite ever finite (and can be differentiated from obliteration at the further end of the continuum), it is devastation that becomes employed as a political tool, performs as rhetorical playground, as data to be calculated, is objectified into things to be traded (such as toxic waste or permissions to do climate damage), and is regarded as something from which some cynical value can still be extracted solely by the differentials to be codified and gamed in the midst of their becoming. In this regard at least, devastations characterize much contemporary thinking around ecologies.

Deleuze proposes, instead of the viewpoint of the beautiful soul, a Nietzschean affirmation of aggression and selection played out in differential terms, which may involve a sophisticated ability to work with what Bernard Stiegler calls the *metis* of politics, an art of war that is neither walled off from metaphysics nor naturalized by it.[34] Michel Serres writes on *metis* as an art of working in and manipulating the capacities of measurement as deployed in the determination of land use, something that intimately couples ecological processes in agriculture and flooding (such as in the assignment of fields along the Nile) with the abstract forms of mathematics.[35] Perhaps articulating something of this condition, there is a certain confluence of operations between warfare, or the exercise of violent power (with or without the monopolies of the state), and the dimensions of the unthinkable manifest in the potential for abstract empiricism. Both are condemned to operations within certain kinds of fog.[36] As such, the recognition of devastations is rendered partial by their inexplicable sense—and hence require a cunning of the kind that *metis* offers. In a manner related to the distinction between the climate and the weather as operative at different though interlinked scales, the ecological aesthetic and medial dimensions of the condition of devastation are significant and yet difficult to recognize as they can be folded within various rationalities and shielded by epistemologies. One kind of devastation is certainly an occurrence without an ostensible aesthetics, in the narrow sense of the term concerned with the sensual and perceptual as such, since there is nothing left to sense them. Such sensing unites the question of the thinking subject or sensitive entity with that of the empirical, the sensible, and the aesthetic as well as with that of occasions where devastation can be deployed as a force. While an aesthetic event ultimately requires no witness, the question of sensing, and witnessing, in this sense is important since it allows for variations to be introduced in the transmission of power.

The legendary destruction of Carthage is one such occasion: like a curse extending to the seventh son of the seventh son, it is one that outlasts our capacity to imagine or to remember it, since by the time such a curse is half-done, its root is forgotten. The story of Carthage was that it was destroyed by the Roman army of the Third Punic War and then broken down, brick by brick, with even the ruins ruined—not, as Alfred Jarry would have it, by making beautiful new buildings from them but by an irrevocable and omnipotent dismantling, and the land being ploughed over with salt, rendering it forever unfarmable.[37] Yet the devastation of Carthage as a site for human life, at least in terms of the poisoning of the land with salt, turns out not to have occurred. The historian Appian's description of the annihilation of the city, in revenge for the victories of Hannibal, nevertheless makes clear the Romans' aim of total obliteration.[38] No one is left to recall the life of the city or what it was like to be its victim, yet there are some grounds, it appears, for its history, since at least it was written. This operation on memory is part of the condition of devastation. The problem with thinking about devastation is multiple. Not only does such an enquiry occasion the problem of the witness—what, if anything, remains to constitute a sense of an account—but also, in understanding the becoming of nothing or of a radical change, how can such phenomena be recognized, if at all, and how is knowledge about them to be produced?

What are the means to speak of the becoming of different kinds of devastations, of blossomings that obliterate? Some, indeed most, things cannot be known by organized forms of knowledge because there are not only so many of them but also due to the problem of scaling second-order, observational knowledge: that of coming up with the techniques of inference, hypothesis, experiment, and modeling, among others. This condition, in turn, causes problems of proof, leaving fractional gaps of doubt available to exploit by those with an interest in maintaining or widening it. In order to close down the operative parameters, science produces instruments, methods, and practices that fillet reality for its juicy bits, taking part at times in this systemic occlusion, and at other times articulating fundamental conditions of multiscalar interrelations. Devastations operate in the condition that Rachel Carson describes in *Silent Spring* in the following way: "Seldom if ever does nature operate in closed and separate compartments."[39] A characteristic mode of devastation, for instance, is that found in the exponential increase in concentrations of poisonous chemicals as they move through a food chain. Samples of poisoned predator and prey species can be subject to biopsies but, echoing the relation between species and individual, not the entirety of the population concerned.

The concept of the witness endows sensing with primary importance. It is not solely about thinking (philosophy) or measuring (science), or gathering and giving evidence. Witnessing unites sense with memory, where evidence rests in bearing witness—a process that can be performed by a subject as well as by an object: a stone, a log. In turn, both science and poetry unite in attempting to elicit modes of witnessing, or better, chains of witnessing, from an event; to a registration of a change in certain molecules, substances, or capacities; to an instrument that is sensitive to them; to a mode of description and comparison that is adequate to them; or to their translation. Yet, as has been described in numerous ways, to witness is also to act.

If to witness is to act, what are the stakes of such a thing, and how are their perceptions logged, turned into memory at the same time as projecting outward? Devastations move between ecological and political scales and across standard notions of both object and system. Devastations can be seen as the process of desertification of the actual-virtual dyad. As such, can they be seen in terms of tabulations of positive and negative, of goods and of bads? Not solely, although, obviously, they may be read off and experienced as such, given certain initiatives and certain organizations (neural, institutional, physiological) of witnessing. The perspectival limits such tables imply certainly can be brought into operation at certain scales, with devastations transductable into their formats from particular perspectives, but such formats can be rather moot at others. The problems of mediation, intellection, and perception shift, combine, and re-sort in the complexity of an ill-becoming.

Can such tabulations ever be capacious enough? Ecology is intimate to humans in every conceivable manner and indeed composes them over both evolutionary time and the life span of an individual, but it is also the condition in which they find themselves stuck. There is a certain degree of intolerability to the finitude of a planet, particularly one in which climate damage has become a form of both political and military gaming field, one operated upon largely by an infuriating indifference that is voluminous in its churning of its own impressive incapacity to act. One of the conditions then of the current sense of devastation is a generalized claustrophobia produced across the immensity of the earth as it hangs amid this roiling fog of a climate.

Part of this claustrophobia is a sense of strife turned, against its nature, into a force of conjoinment. Disjunctive synthesis overrides the friend/enemy distinction when it comes to the composition of the atmosphere.

Yet, as a Peoples' Liberation Army strategy document from the last decade titled "Unrestricted Warfare" noted, furthering Carl von Clausewitz, ecology has become a means of waging war—one unlimited in its scope.[40] Perhaps it is this systemic factor that is becoming significant in the present era. The ineptitude of established political or economic forms finds itself mobilized as their primary delegated means to imagine a clever exploitation of the situation, aping *metis*. The means by which this war is to be fought are in the processes of figuring themselves out and are to be found in the domains of energy and fuels, water and pollutants, and the morphological manipulation of terrains: such as that of northern coastlines via ice melts, the flooding of low-lying countries, and several other means, such as the ruination of fertile soil.[41] Following the consideration of their strategic usage and the problems associated with them, leading to the adoption of the full range of both negligence and opportunism at the level of states' reactions to ecological crises, devastations also impose particular kinds of conditions for knowledge about them in terms of the kinds of cunning required for their exploitation.

As with the case of Carthage, devastations are, among other things, an operative component in systems of war. The capacity for them, the carelessness with which they are handled or flaunted, and the opacity with which they are left as the world moves on characterize their strategic value. Such a form of becoming of munitions is active with the residual and freshly seeded crops of land mines, chemical and biological weapons, cluster bombs, and nuclear weapons and their equivalents in industrial accidents: a constituent part of modern warfare in both its implemented and threatened states, as part of its operation as calculus, trauma, and frenzy. Each of these forms of weapon gains part of its power from the violation of ethics that they imply, and also for the unknowability of the violation of the future that their use unleashes. The calculation of the cost-benefit ratio of land mines, for instance, sees them deployed widely and rapidly as a means of asserting control over a territory, making it impassable. The condition of wild-seeding of such weapons sees them left in the ground for decades, a momentary tactical or strategic advantage lasting in swathes of unfarmable, impassable land. The deployment of nuclear weapons triggers the exercise of devastations as the actual settles into a state of strategically engineered "irresolvability."

The *political plastic*, as Eyal Weizman calls it, is generated out of the interaction of forces, potentials, and the affordances of entities such as laws, turned into calculuses of the permissible and the bendable, the reach of weapons systems, landscape measures, and also out of potentials of retaliation, of destruction and modelizations, and the analyses of such.[42] Indeed, the international history of the Cold War could be written through the interlocking systems for devastation and the mechanisms for making them implicit but calculable, known but ineffable, operative yet unused.[43] What is interesting about this particular sort of plasticity is that, like that discussed by Malabou in relation to neurobiology but operative at different scales, it has its limits—but these are only discovered or used as momentary tropes within a larger set of fixings and changes in a sort of parametric emergence of a situation out of things without measure. The question of devastation in relation to sense, witness, and warfare can be seen as a question of measurement and is treated with a remedy of calculability. Calculation of the unknown extinction is one artifact of this condition and should perhaps be recognized as crystallizing the dire conditions of devastation in relation to the problematics of knowledge.[44] These kinds of tryouts of little devastations—calculated and modeled diminutions, a training and development system anticipating the larger-scale devastations to come—exist in a tensile axis composed between the presupposition or impossibility of a condition of general calculability. This axis is in turn a response, or a conceptual pair, to the condition of irresolvability.

What is notable, however, about the question of devastation is that the techniques of observation that attempt to capture its characteristics proliferate according to context but, as methods, need to be repeatable in order to gain greater traction on the problem. But since devastations operate often by the becoming of loss as well as the growth of something unknown, they are paradoxical, because what we are able to recognize of them is both a form of presence and an absence, producing a version of the logical problem of the evidence of absence. How do you prove the dissipation of the virtual? You may need a vivid imagination, or perhaps you may simply need to be glacially cold, painstaking. Perhaps indeed the latter, since devastation is in a certain sense the knife-edge upon which present social forms find their seat, and yet the knife-edge that is less than infinitely expandable.

DUMPS IN BODIES

Certain ideas about nature have a tendential form of operation in that what is sectored off as nature becomes nonconceptual, passive, or overwhelming. There is, for instance, a certain overlap between emphasizing the awesomeness and unknowability of the sublime and the idea that nature can absorb all that is thrown into it.[45] The mighty and eternally flowing river Yangtze makes a perfect chemical dump. The North Sea can be overfished, it is imagined, in perpetuity, while nations bicker about their territories. The steppe stretches so far that it can absorb anything. Overcome with the power of nature, coupled with the operations of other ineffable mechanisms that condition knowledge (as discussed above, such as markets and their convexly mirrored versions produced in externalities and the unowned and unowned-up-to commons), the unknown is used as a dumping ground. If we are to think of the media of ecology, then Earth is a means of mediation, a pretext for deferral, a hyperabsorbent diaper or nappy for an incontinent humanity.

As well as dumps in seas and in landscapes, there are other kinds of volume being exuded but into the flesh of human bodies: surplus production that must be forever devoured, regurgitated, chewed, and gorged. As is well known, there is an epidemic of obesity in the world in which human bodies become the sites for the dumping of certain kinds of surplus. Obesity itself becomes an ecological crisis since it involves an increase in the volume of gross human biomass.[46] Based on 2005 figures from the World Health Organization, increasing population fatness is projected as having the same implications for world food energy demands as an extra half a billion people living on Earth—one needs more food to sustain higher weight.

Causes are multiple and of varying kind and interpretation. Aside from the variable genetic predispositions of individuals, this situation of growth is characterized by changes in food and access to quasi foods, increased mediation of food into a signature of surplus that is yet unmatched by *Homo sapiens*' ability to devour and offload it, lack of food and abundance of access to foods with high calorific value and lack of other kinds of nutrition, and the persistence of a kind of body evolved in the context of hunter-gatherer forms of life into historical conditions more suited to species able to benefit from high quantities of sugar. Seen solely from this perspective,

Western civilization with its ramping up of sugar in the average diet is more suitable for inhabitation by generally more physiologically simple species, such as slimes, bacteria, and algae that are more directly able to translate such abundance into reproductive activity. This in turn can be figured as a form of devastation. What we find with obesity, however, is that more structurally complex organisms can be said to internalize and mediate certain devastations at the same time as they are the grounds of them. This condition of the internalization and mediation of economically and politically expedient surplus is what characterizes the obesity epidemic as a peculiarly contemporary devastation.

Obesity has many factors, but they are conjoined in the particularities of the way in which humans articulate more general biological characteristics. Food is mediated within the body by hormones, particularly the homeostatic factors, such as ghrelin, which helps signal hunger, and leptin, which signals the state of satiety. These in turn may interact with dopamine, released by the ingestion of food found to be delicious yet decreasing in the amount yielded the more is consumed—a condition, in turn, amplified in the obese, who have less dopamine receptors, less capable of producing the required effect. Within the body, multiple other systems are involved, such as the activity of fat cells, which are not simple warehouses for energy but are also productive—generating fatty acids and hormones, among other things.[47] In turn, conditions such as diabetes, cancer, stroke, liver failure, and heart disease also have their particular capacities of formation in relation to such factors.

As Guattari notes, systems of endocrine regulation may hold "a determining place at the heart of assemblages," giving a particular stubbornness or lubricious ease of implementation to certain social configurations.[48] Such capacities of the body can be hooked into by particular substances and the assemblages around them, generating a virulent *conatus* between an agricultural policy, political tactics, human appetites, and the condition of obesity. Such actors inevitably bump up against one of their mediating components—the idea of human subjectivity and what is taken to compose it.

Lauren Berlant has deftly argued that prevalent systems of the understanding of sovereignty as applied to peoples' negotiations of "ordinary living" in the overdeveloped world is effectively a category error, one that filters people into a double bind.[49] There is a mismatch between the attribution of guilt to those who are inefficient calorific self-managers in a condition in which sovereignty is culturally required but politically, economically, or distributively ephemeralized or removed as a possibility. The question of whether a given activity—such as eating or drinking a certain composite of things—can be described within particular discursive frameworks as good or bad, conformist or resistant, drifts and congeals, like the self-organizing of a nonpolity that accompanies and acts as guarantor to the commons of pollutants, toward the vanishingly indifferent: indifferent, that is, except that—partially by means of such a negligible status—it is something that actively textures and differentiates the multiscalar politics of everyday life, leaching into spatial politics, interspecies and class relations, and the possibilities for the invention and sustention of new gustatory and organismic pleasures.[50]

Berlant's astute biopolitical reading of the condition of obesity and the multiple discourses feeding it can be complemented by one of ecological ethico-aesthetics. One of the lines into this condition can be traced by following the vegetal biopolitics of corn. Richard Nixon's need for the support of farmers leading toward the 1972 election concentrated the farming practices of the Midwest around corn by providing federal funds for those growing the crop. The achieved surfeit of corn required its uses, aside from feeding to a glut of cattle, leading especially to high-fructose corn syrup (HFCS) entering, or being dumped into, the American, and thus global, diet.[51] This had the benefit of lowering the market price of certain kinds of food in a more general condition of inflation. Once a product and a market was created, it persisted, as did the federal subsidies. HFCS is found in soft drinks, processed meat products, bread, sauces, cereals, and many other food and foodlike substances. In those in which extensive processing has decreased flavor or in substances used as food or quasi food in which flavor is not naturally occurring, it is useful as an additive. Eating or drinking HFCS represses leptin, and so diminishes the eater's capacity to recognize that they are full or sated. Such a process has no need to occur with the full knowledge of what is occurring in any of the participant humans, nor in the agencies, markets, instruments, glands, intestines, brains, plants, policies, political intrigues, taste organs, or other entities involved. As Berlant notes, "The image of obesity seen as a biopolitical event needs to be separated from eating as a phenomenological act and from food as a space of expression as well as nourishment."[52] Such a conjunction is sorted, amplified, ablated, contused, digested, and stored by the interactions of the predilections, intents, and desires of the particular systems brought together in the ensemble; separating their constituent aspects becomes an analytical work that requires ecological acuity.

Human bodies are places for regular substance panics (such as those associated with acrylamides, saturated fats, Bisphenol, plasticizers, and microplastics, etc.). Their ecologies are combinations of complex chains of media, from the instruments and recording devices of labs, the persuasion mechanisms and institutions to which they are attached, to those of televisions and the mechanisms falling under the scrutiny of the discipline of communications, and those ecologies tangled and forged inside organs and food and logistics systems. Characteristic of these is the reaction to the discovery that human milk becomes toxic when it concentrates chemicals such as PCBs stored in the mother's adipose tissue throughout her life.[53] (Obesity is perhaps one of the necessary requirements of contemporary life in that we need sufficient space to warehouse and thin out all the toxic chemicals we are exposed to as part of a distributed and disavowed necro-commons.) The agency of such chemicals, residues of mindlessness toward matter, turns the body inside out, rendering moot the scale to which it is most fundamental and to which agency can be attributed. Here, the question of movements of dissipation and concentration of chemicals in a dispersed set of states and sites within an ecology becomes crucial (whether such materials are ideally to be recycled or warded off, or digested with some dose of resignation or glee) and ties in with the question of energy—how much energy is needed to gather all that must be recycled, or to recoup all the matter that has spilled into a condition in which it is poison.[54] Beyond a certain point, which is not always so far, there is a devastating becoming that makes certain kinds of known lives untenable.

Devastation thus has no particular scale, no necessary consistency or cadence. It may be found among lakes and amid organs, in the expanding edge conditions of deserts and in the sifting out of the conditions of possibility for species. Equally, it is found in the way in which the means to recognize, to understand, and to invent the means of acting in conditions of ecological complexity are vitiated or go unworked. It is off-loaded as an externality or as a necrotic commons sustaining null becomings. The textures of this dynamic provide incipient and full-frontal conditioning to contemporary politics, saturation economies, and modalities of conflict as well as provide exemplary processes for subject formation. Finding the means of naming, tracing, and disaggregating devastations means also to find ways of living in absences without the salving benefit of any finality, thickening, in turn, potential preludes to other, more lauded dynamics of becoming. This is likely to be difficult. Indeed, to do so may partly entail coming to experience, or finding means to distill, the mutant form of Spinozan joy known as anguish.

Anguish

◆◆

A young girl, at most five years of age, gives an elderly man a jug of water laced with rat poison. Drinking the water kills him. The man and the girl are neighbors sharing a labyrinthine flat with a huge balcony, one overshadowed by trees and sprinkled with fallen apples. The balcony becomes the scene of the murder. The man is wheelchair bound and looks after the girl while her mother is away at work. The mother provides supplies to the old man in exchange for childcare. The man resents the girl's mother. The girl resents being stuck at home.

Though it is a conscious act on her behalf, it is also clearly not an act of full consciousness; it is neither instinctual nor unclouded goal-seeking behavior. The scene is not wrong, bad, or even sad. There is nothing about it to be stated other than that this event is enunciated as an activity of life, a coming together of various conditions and elements, similarly to how hail happens through the interaction of water and cold layers of air, though much less systematic or prone to repetition: the girl's desire to go out and play; the chance occurrence of rat infestation and therefore poison lying around; mutual animosity; the request for a glass of water from the old man; the absence, we may guess, of certain taboos. The girl dresses and undresses, makes faces in the mirror, hits dolls—she is bored. She forgets and remembers again what she was about to do, collecting poison from various rattraps into the glass awkwardly and in a manner of absent-minded playfulness. The water in the glass is murky: propitiously concealing poison, she changes cups, pouring their contents into an enameled mug. There is inventiveness, sharpness, and jocularity—all in all, a vitality of matter that proceeds in this particular way, which ends one shabby life without posing a question, whether of value, truth, or evolutionary advantage.

This is the last segment of *Three Stories* (1997), a film by director Kira Muratova (1934–2018).[1] Born in Bessarabia (at the time Romania, then the Soviet Union, modern-day Moldova), Muratova had a long, though often obstructed, career in Soviet cinema. She made her films in the Russian language and lived in Ukraine's Odessa. Muratova was "interested" in death. She was often accused of misanthropy. Among the scenes in her films, there is an ability to look where others might turn away. In *The Asthenic Syndrome* (1989), there is a five-minute-long scene of stray dogs in a dog pound, many of them sick and disfigured, in death angst. They are about to be killed. The opening scene of the last segment of *Three Stories* is of a black kitten tearing apart a stolen dead chicken on a high brick wall. The kitten is trying to hold its own balance on the wall and that of the sprawling chicken, while biting at it. The scene is both innocent and troubling, something to be repeated later with the child and the old man. The kitten is small and pretty but angry and unpleasant; the chicken looks disgusting, and the efforts that go into holding balance while half-eating, half-trying to get away with the body fill the scene with a macabre character. Joy, innocence, and death meet here too as they often do in Muratova's films. Muratova does not judge or mawkishly gaze at the atrocities of life; rather, she displays certain movements of vitality in an attempt to attune to specific kinds of experience and give them aesthetic form. It is through her work that we can make an entrance to the sensory, sentient, and reflexive experience around which this chapter forms, that of anguish.

The difficulty of naming anguish, similarly to that of naming devastation, lies in the scarcity of readily available ethico-aesthetic means of production: some old linguistic, aesthetic forms and devices are frazzled; others are parasitized by the sprawling modalities of contemporary capitalism; and the remaining ones are marred by a mixture of deconditioned but lingering modern narratives of subjecthood. The condition of today is that of being partially mute: there are not enough words to speak about things we want to discuss. The term we settle on here, anguish, is unwashed. As a notion, it is too human-centered and firmly associated with individual depression and dysfunction; in the optimistically geared English of today, its sets of connections lead us astray. Yet we propose to ease it out from its legacy of associations and revivify it. This effort is part of our project to develop a vocabulary for ethico-aesthetics that is not simply human but is distributed in wider ecologies. The terms "fatalism" and "glory" developed in the "Plant" chapter are related parts of this project.

It is essential to delineate anguish from trauma, shock, finitude, and negativity. The most familiar way of thinking anguish is through its instrumentalization as depression with its subsequent identification as a medical condition (to be treated with antidepressants).[2] Anguish is also an object of psychoanalysis (as explored in Julia Kristeva's *Black Sun*) and has a history in the disguise of melancholy, as a quality of temperament and also of intellectual character (that is how Susan Sontag wrote about Walter Benjamin).[3] There are various terms for anguish, including distress, despair, pain, chagrin, grief, and angst, each of which had their own historical moment and epitomizing character. Anguish is related to tragedy and suicide but also to episodic and nondecisive suffering. The above can be crudely summarized into a few kinds of anguish: a mode of experiencing and a character of thinking, an eventful consequence of something, a phenomenology of consciousness (Sartre's existentialism), a biological condition, and a biopolitical object to be managed, treated, gained from, and exorcized.

In a certain way, anguish is also absent from the vitalist philosophy of Deleuze and Guattari. As is well known, Guattari suffered from depression, but he chose not to write about it. In fact, especially in the 1980s, Guattari writes himself out of depression, his work being to claim the world affirmatively.[4] Conversely, Kristeva writes explicitly about her depression, and it comes entangled with an abyss that carries paralysis and inability to work (an exact opposite of affirmation through production). In this sense, it shares a legacy with "ugly feelings" as conceptualized by Sianne Ngai, as a diminution in agency.[5] Lauren Berlant also works on reevaluating human agency, in her discussion of "slow death," against habitual and domineering imaginaries of sovereign subjecthood and decisional capacity.[6] Here, singularized depression, diminution, and slow assault extend unnoticed to populations, their organs and dreams, while sprouting negative feelings that help discriminate against self and others.

While the notion of anguish this chapter aims to develop is not isolated from such conceptualizations, it is not exclusively on the side of the ugly, cruel, or heftily bad. Slow, but at times fast, anguish is shared in ecologies that do not necessarily experience diminution in agency; and if a redistribution of agency occurs, it is not uniform. In Chernobyl's long devastation, it is not only isotopes of plutonium and americium that thrive but also animals. Because anguish is a mode of experience that extends subjective agency to nonhumans and recognizes the ethico-aesthetic sensorium of ecologies (related to what Gregory Bateson called the ecology of mind), its complex causality and uneven distribution of gains to participating parties but also its metaphysical openness and undecidability do not square exclusively with negativity, imagined as exhaustion through diminution or collapse.[7] While anguish can be an ethico-aesthetic experience of the depletion of the world or of the parching of the virtual, it is not limited to devastation. Rather, anguish is an ethico-aesthetic sensibility forged amid abundant productivity that may also be menacing. It remains important to stress that while the materiality of anguish does not necessarily register at the level of endorphin inhibition, or of black or yellow bile, and does not necessarily involve human suffering, we must not negate anguish by flipping it over into a flat affirmation of all activities of life.

The related Russian word *toska* is another terminological possibility. Vladimir Nabokov said about *toska*: "No single word in English renders all the shades of *toska*. At its deepest and most painful, it is a sensation of great spiritual anguish, often without any specific cause. At less morbid levels it is a dull ache of the soul, a longing with nothing to long for, a sick pining, a vague restlessness, mental throes, yearning. In particular cases it may be the desire for somebody or something specific, nostalgia, love-sickness. At the lowest level it grades into ennui, boredom, *skuka*."[8] *Toska*, while being felt individually, is an aesthetic concept that is systemic, which means that one gets exercised in experiencing the world in its ways. *Toska* is a sensation and yet a concept, a conceptual feeling. As an aesthetic prefiguration, it is productive of reality as well as being reflexive of it: so while we experience *toska* with body and soul, it also dwells on the levels of abstraction. It is specifically effective in cultural production, and its place of residence includes literature and film, and yet it is also generative of and inculcated at the scales of habits, memories, decisions, forms of life, and social formations. The word itself, though, is not ideally suited either: somewhat of the nineteenth century, it gleams with shades of grandeur, metalosses, and yearning for meaning that necessitates a great price.

While indebted to *toska*'s multiplex character, anguish in this book is an ethico-aesthetic dimension of being, which is not necessarily concerned with the problematic of the human. This figure of anguish proposed here is an attempt to "modernize" *toska* and "dehumanize" anguish, while describing certain characteristics of aesthetic experience fit for multidimensional things stuck in processes of devastation or other forms of alteration of future unfoldings. While its legacy is often linked to guilt or to fault—products of a desire to explain suffering in terms of clear causation—anguish is not limited to one species or particular kinds of things. As it goes ecological, anguish loses linearity, habitual causalities, and even the hopeful project of progress; shedding teleology, it loses a link to automatism and incapacity.

If anguish is set as a mode of experience—a sensation and reflection—it can habitually assume a subject, even if it is a fuzzy, diffused subject. This is because the movement of anguish is to condense onto or to coalesce differentially upon a scene or a specific composition of things rather than be equally distributed or uniformly overpowering.

Just as devastation can exist outside of witnessing or knowledge about it, its sensory reflexive capacity can become one of anguish. This capacity does not necessarily presuppose a consciousness but an ability to be affected and openness to radical alteration of the future in the shearing and torsions of scales. Anguish thus shares characteristics with Spinozan passive passions, and in cases where it pairs up with devastation, it may become a way of experiencing the previously discussed "destructive plasticity." The possibility of destruction is a constant ontological option and yet always remains an unanticipated accident, thus logically and biologically available and forever haphazard.[9] Malabou's ontological plasticity is a mode to be affected, both positively and otherwise—for instance, with brain injury, when one's existence is modified absolutely and a tendency to preserve one's being and experience subjective continuity is radically altered.[10] It can be fruitful to interpret Muratova's scenes of the murder mentioned at the beginning of the chapter in terms of a haphazard and yet always ontologically present possibility of destruction. Malabou's ontological plasticity is the premise of the old man's death in Muratova's film.

The accident of his destruction is momentary and contingent. Yet other participants of this event partake in its anguish in different ways, and the consequences of this drawn-out event are less destructive for some of them in terms of their plasticity than for the others. Human beings enter anguish and become part of it as much as they are parts of other processes and forces; anguish does not originate in the human or become locked in it but traverses it. Anguish extends before and after the event, its time oscillating between episodic and intensive time, including both. An individual line of life may shatter, but dynamically unfolding engagements of things and of their conglomerations will obscure any simple causality. The perspectives offered to the viewer are multiple, including those of an innocently murderous child, with the innocent kitten, innocent dead chicken, and innocent apples scattered on the floor, inhabiting anguish at different tempos and intensities.

The subject in this anguish can differ from vertebrate to insect, and even a cell, drawing in all instances of individuation but also emphasizing concatenations of things and processes. Apples, plates, and tables are drawn into and participate in an ecology of anguish. A pile of powder or a chicken's corpse, a fruit or a wall inhabit and experience anguish differently from the way humans are traditionally expected to, yet, while it may fit at times, it is key not to place anguish at an anthropocentric level of transcendence. This is not out of any good manners but simply out of the need to make an enquiry adequate to the factors that produce its time. While anguish combines the orders of affection and abstraction, it does so without posing the latter as the sole capacity of human reason. Here, the planet consists of orders of abstraction all the way up—and down.

In some way, anguish is conceptually close to Whitehead's formulation of feelings, as something that exists at physical and conceptual scales and operates amid subjects of different scales and orders. Whitehead says, "The whole universe consists of elements disclosed in the analysis of the experiences of subjects. . . . Apart from the experiences of subjects there is nothing, nothing, nothing."[11] Needless to say, such subjects' experiences do not need to be that of the mind any more than of a quark or a rissole. Anguish coalesces upon ecologies of subjectivities of various kinds, drawn into devastation or other operations on the virtual. But there are important differences between anguish and Whitehead's feelings that are, similarly to

Spinoza's affects, relations and interactions that operate cosmologically. Despite being described as sensation and popularized as affect, such feelings are modes of action and construction. In Whitehead, feelings feel the data and select and change them as part of the process of prehension (coming into being as actual occasion). Prehension itself is a process of feeling. And feeling the data massages the figures, sometimes with claws. Here, in his terminology, physical prehension is feeling and interaction between actual entities, and conceptual prehension is interaction between actual entities and eternal events. Feelings thus enable things, constructing the world in both material and abstract terms. Anguish, however, is not itself a mode of interaction or construction. Even as an abstraction, anguish does not select and change things. As an aesthetic and generative mode of experience, anguish does not have a prescribed and foundational cosmological function. Unlike devastation, anguish is not a mode of operation on the virtual but rather an undergoing in and of the world that is stretched on an axis of devastation or other mode of becoming, rending them in or across scales. As such, anguish may come into being as collateral damage, as a mode of self-recognition of a state of being. Anguish describes ethico-aesthetic sensibilities forged amid negative or nebulous, opaque productivity in ways that are neither anthropocentric nor dialectic (will not result in catharsis). It is neither explicatorily exhaustible through measurement nor abstracted without a material expression.

Anguish is rather passive, in that it is not interaction itself nor the aesthetic quality of a unique interaction, some kind of sensitive representation of a bad occurrence. It can be imagined as a conceptual mode of sensation, experienced in the betweenness of subjectivities. Anguish is not fixated onto a locus but arises in multiple intersecting loci diffused through ecologies. Anguish does not preexist things, waiting to creep up on them or for them to embody it as an ideal, but is produced in processes of living; experienced subjectively, its subjects are of multiple kinds. The operation of thinking anguish is thus difficult and involves the doubling and trebling of points of view, following forking legacies, and going against their lead. It is an aesthetic presence of the world but not only a presence available to or produced by certain privileged subjects. It is a figure coalescing upon and experienced by meshworks of subjectivities but without strict confines of selfhood, minds, or nervous systems.

What kinds of dispositions then take on anguish? In *Without Criteria*, Steven Shaviro offers an interpretation of Whitehead's proposition of the "beautiful": one that moves between the image of judgment in Kant and the relation in Deleuze and Guattari. Shaviro suggests that for Whitehead, the feeling of the beautiful happens to subjects, which is what the world tends to coalesce upon. Beauty here is an aesthetic event of attraction outside good and evil or true and false. Such beauty forms the basis for a Whitehead-inspired rereading of the discussion of the parallel evolution of the orchid and the wasp in Deleuze and Guattari's *A Thousand Plateaus*. Among a number of species of orchid in which the flower mimics a female bee or wasp, drawing males in to carry out pollination, the flower is found beautiful by the wasp because "the orchid 'adapts' itself to the way the wasp apprehends it."[12] Various species of orchid benefit from the luck of an uncanny overlap through the evolution of patches of blue that resemble wings, sometimes a furlike penumbra to petals edging what seems to be an abdomen, and an aroma that resembles a female insect. They remain indifferent to each other and yet "interested" through the mutual yet asymmetrical benefit of pollination and implied sex.[13] Deleuze and Guattari, Shaviro suggests, use the idea of the beautiful to interpret such relations of what and how things become but only in relation to each other. One can easily imagine that once the dynamics characteristic of such structural couplings move into more than two bodies, concatenations of interest and multiple chains of indifferent mutual benefit and enchantment also arise. By such means, anguish may also proliferate.

Anguish's terms of relations between things and of their mutual becoming are another such product of mutual "interest" but are of a different kind. Here, parallel adaptation does not occur and no relationships of beauty spring up. What happens is at times logical but nevertheless obscured by a quality of indirection—absurd and unexpected destitution, a destruction or subordination. These things have not evolved toward each other and are not quite apprehensive of each other. Neither are they necessarily engaged in relations of mutual animosity or in a fight. Brought into relation, yet unrelated, their encounters are fortuitous, albeit consequential. As when an animal stamps upon an orchid, and a wasp is killed by out-of-season snow, anguish is not a product of encounters that are absolutely random yet their undreamed logic changes the unfolding of the future.

Anguish can be thought of as an alternative, though not an opposite, not a geometrically "equal" partner, to Deleuzian bliss, something discussed in detail in the second part of this chapter. In *Pure Immanence*, Deleuze presents jouissance, or bliss, as an experience of vital matter.[14] What is this bliss? It is expressivity, affirmation, the creativity of vital processes of life, both immanent and transcendent. We ask instead, as it becomes urgent, what if and when the bliss is not blissful? What if this is an anguished bliss, a vital joy of *toska*, a bleak joy?

Anguish is a result of annihilatory vitalism and unbearable productivity—in direct terms and in chains of consequences that may follow. Anguish may arise as a character of productive and vital forces cutting across different scales, not united by the direct or indirect ecological commitment of one element and scale to another but through their jarring and dissonant unison spread across times and spaces. On the other hand, anguish can be experienced in perfectly logical and tightly knit systems: families, platoons, selves, forests, ponds, sewers—there are multiple kinds of constitution of anguish. In what follows, we trace the question of anguish in previous historical periods, through Nietzsche and Deleuze, and in relation to the formation of value, causation, and modes of resolution. We describe some insights into anguish from Varlam Shalamov and Kira Muratova. The chapter then proceeds to inquire into some characters and modes of action that anguish can generate, before considering its place and possibility in the vitalist ontology of immanence. The last part of the chapter talks about the modes of distribution of anguish and anguished apparatuses, whereupon we offer some unhopeful last remarks.

GENEALOGIES OF ANGUISH

Theories of decisional capacity in humans have stressed the importance and centrality of emotion, including "negative" emotions, to decision-making capability. Emotion is also central to forming value and values—hence its pockets are picked by consultants and cognition-tweakers for means of upping the ante in performance. Anguish is traditionally fundamentally related to meaning and value: mostly, of human life in relation to its pain and dreariness. For a long time to inquire into anguish meant to inquire into the purpose of life with a question: is life worth living, if it is so full of suffering and pain? It is not only the meaning of a loss and the loss of meaning but causality and teleology that are packaged into this question.

The question has been tackled by numerous philosophers in, as Clément Rosset puts it, an attempt to think the "worst."[15] As Deleuze argues, that is the question that Nietzsche inquires about when he writes *The Birth of Tragedy*. As often attested, Nietzsche was still under the influence of Schopenhauer, a great thinker of unhappiness, when writing *The Birth of Tragedy*; however, the concepts and disruptions that he brought into twentieth-century philosophy are already there and it is particularly the book's interpretation by Deleuze that we follow below.

If, according to Deleuze, in Schopenhauer, will was the external force that made any human happiness impossible and it was the nature of the universe that defied continuous human happiness, the project of Nietzsche was to defy the finality of any external force. For Deleuze, Nietzsche's main task is philosophy of value, as a critique of values. Values, thus, should not be derived from a universal principle or from a local one, in manner of a consequence. Values can be worked out as critiques, an active creation: "divine wickedness without which perfection can not be imagined."[16]

The Nietzschean critique of a tradition of dealing with pain that valorizes it and imbues it with meaning is the critique of a few ways of managing what we might call an ancestor of anguish. One method, applied by Socrates, was the force of abstraction, a reliance on external sources for making sense of what happens on Earth. This reached its highest point in Christianity but also in Hegelian thought, where the redemption that relies on the functions of guilt (internal causality) and fault (external causality) serve as the valorizing meaning-making abstract machine.

Nietzsche attacks dialectical philosophy and the process of valorization of suffering and sadness as it relies on the idea of positivity, itself a product of negation.[17] Nietzsche is vehemently against a still prevalent tendency to produce meanings and seek causality (one suffered—one has been saved; one suffered—one learned) where there is none. His critique is based on the idea that if one assigns value to suffering without trying to resist it, suffering becomes a cruel must for all humanity without a possibility of opting out.

The justification of suffering, or at least its valorization as a purgative activity, is very much present in the writing of Alexander Solzhenitsyn.

That is why Varlam Shalamov's writing about prison camps is often opposed to that of Solzhenitsyn. Both draw on experience, but for Shalamov, the experience of a camp is unjustifiable: those who survived did so not because they were stronger—whether physically, mentally, or spiritually—but for *no reason* at all. There is no reason why one person survives and another does not. There is nothing to learn. One cannot make sense of or see a purpose of being in a camp. There is no value to be drawn from it, no positivity to be established as a counterbalance to its negativity.

While both Nietzsche and Shalamov refuse the attribution of redemptive qualities to suffering, for Nietzsche, going further, it needs to be embraced. Yet for Shalamov, while there is nothing camps can teach us, and this experience has no value, it is better never to have it. There is no point embracing it. Shalamov writes about the administration of a concoction made of the needles and twigs of the creeping cedar (*Pinus pumila*), supposedly rich in vitamins, to prisoners: "At the time many drank the stinking abomination, went away spitting, but eventually recovered from scurvy. Or they didn't recover. Or they didn't drink it and recovered anyway."[18] This is Shalamov's ontology: no bliss, no affirmation, no synthesis, no conclusion, no causal logic except for combinatorics. "We all understood that we could only survive accidentally."[19] A person in this story, who is sentenced to ten years in a camp, hangs himself—without a rope, just by placing himself in a branching point of a tree. Shalamov writes: "If he were to die now, he thought, how cleverly he would have deceived those who had brought him here. He'd cheat them of ten whole years."[20] This person fooled the authorities by snatching his ten years' sentence from them— by dying. This dreadful logic stands: in such magnitudes of suffering, constructed and maintained by other humans, a clear although vile causality is at work—one that can be described through ideology, war, terror, authority, subservience and capital, all equally meaningless.

While not divorced from such legacies, anguish has no clear causality. It is difficult to establish why or how exactly devastations or other operations on the virtual play out. Anguish is a nonlinearly generated mode of experience. As such, it does not have a clear reason or a registrable set of features; it can go unknown and unnoticed by its observers, witnesses, or participants. Shalamov's horrors are at the far end of the continuum of anguish: of unbearable pain, atrocity. Anguish also plays out quietly, often with delayed registration, arising in a matter-of-fact way without matters or facts, devoid of clear causes, consequences, or purpose.

Nietzsche claims that we would not be able to withstand pure horror; we are able to survive the witnessing of a tragedy only because of its illusory appearance. Turning the question of pain toward that of the capacity of representations to embody it, he advocates for a pre-individual, Dionysian state that can be brought about by witnessing the horrible (and pleasurable).[21] The question of the intolerable, unbearable, and unrepresentable is another edge of anguish. The elusive character of anguish lies in the fact that it is not quite a proper tragedy. While anguish can certainly relate to tragedy, it is not a representational means to help deal with the "paralyzing vision of the intolerable." Anguish is not a representational aid or cover. "Proper tragedy" is out there, at the horizon, but anguish arises when it is not fully arrived at, at least not yet, not definitely, or is draining away. In Muratova's films, anguish is not paralyzing and eschews proper tragedy on the ancient Greek scale, even if one thinks of it outside the film form. The different quality of anguish today is due to the confusing kinds of production of life and operations of power, such as irresolvability (explored in the next chapter), in the lack of causation and representation, in the absence of someone to be made accountable for it—as it lies in the state of diffusion, answerable only to bad luck (discussed in the chapter "Luck").

Nietzsche wrote that knowledge of distress characterizes times and periods better than anything else.[22] A previous age, an age of fear, was one where violence had to be inflicted upon oneself to be trained to an environment of pain; in his day, pain was a torture, and as with tragic philosophers, only the thought of pain became utter pain. Shalamov's pain is undoubtedly a mid-twentieth-century knowledge of distress, as repetitive and unique as it can possibly be. And yet even that knowledge has already moved further, and with it, the experience of anguish.

What kind of answer does Nietzsche offer? His answer lies in affirmation. For him, pure joy and the profoundest sense of tragedy are the same. To experience the greatest possible amount of pleasure, one must experience the greatest possible amount of pain. To remove pain means to minimize the capacity for enjoyment. And then, if maximum pain is embraced, its counteracting force can be discovered simultaneously. Nietzsche says it is believed that good elements are "conservative of the species" and evil elements are "detrimental" to them, but evil (pain, violence) can bring favorable circumstances.[23] Nietzsche's maxim is "The remedy for 'the distress' is distress."[24] Deleuze, in turn, writes, "Seeing or inflicting suffering is a structure of life as active life, an active manifestation of life's pain is not an argument against life, but a 'stimulant to life' for life. One cures oneself of pain by infecting the wound."[25]

The tragic here is an aesthetic form of joy, a joy of the multiplicity of life. The pure tragic is affirmative, an "inseparable ecstasy and suffering of human existence."[26] Can everything then become an object of affirmation, of joy? What is purely tragic and how can it be handled? What is this new tragic culture Nietzsche wants to construct? The new tragic culture suggests getting away with the optimism (either Christian, affirmative in the acceptance of God, or Schopenhauer's negation of the will) and beyond the hiding powers of abstraction (Socrates), though involving thought. This tragic does not consist of contradiction and resolution, of life and suffering, destiny and a universal spirit. Nietzschean tragedy is Dionysian, dedicated to the only "suffering and glorious" god, in whom the suffering of individuation is absorbed into original being. We can be defended against despair and lured into life by a force of synthesis of the Apollonian and Dionysian: not a Socratic response of using knowledge to shade away from our destiny into abstraction but staying within the immanent life, joyful in the state of affairs. In Deleuze's interpretation, instead of negation, opposition, and contradiction, Nietzsche introduces difference as an object of affirmation and enjoyment.[27] Nietzsche praises pure irresponsibility as withstanding guilt and fault, and pure affirmation of all becoming in its innocence rather than raving about it being cursed. Deleuze, in turn, condemns anguish as a force of nihilism. In *Pure Immanence,* he writes that anguish, an "uneasiness about life," is an "obscure sense of guilt," whereby the valorization of sad passions and negative sentiments is itself the mystification done by and in favor of nihilism.[28]

What does this mean for anguish? What seems to function as the limit for anguish is not only pain, or atrocity, or unrepresentability, but immanent thought, which excludes or transforms it into something else in order to proceed.

With rare exceptions, anguish is indeed silenced and unseen. It is not put to service, and this is not only because neoliberal regimes block any nihilistic impetus but also because novel forms of anguish that are not yet valorized are bred. Nihilism worked in a Christian model, toward specific ends. Contemporary conditions complicate and convolute nihilistic systems of coordinates, whereby although full negation is often denied, we have not always graduated into the possibility of full affirmation.

Anguish has its own form of energy, its own kind of becoming—not in relation to pleasure or overcoming pain, to negation or redemption. Anguish is not a mere consequence, and while it can manifest as an aesthetic creation, it is also imbedded within a preconscious immanence that does not easily submit to evaluations that align with values: juridical, moral, or economic. Anguish is not tied to purity—nor to nothingness; it is full of force in itself. But we still need to understand what it is productive of, and to question its force, as anguish is stretched out between its poles, oscillating between possible intensities and limits, and trembling in its conjugation of subjects and relations.

DISCONSOLATE AMONGNESS

One way to think about anguish is by accounting for what inhabits and appropriates it. Echoing Foucault's description of power as action on action, Deleuze writes, "The sense of something is the relation to the force that takes possession of it, the value of it is the hierarchy of forces that are expressed in it as a complex phenomenon."[29] One kind of processual force that takes possession, aligns itself, or rides on anguish is individuation.

We can look back at the Russian notion of *toska* to think anguish as part of individuating force. To be in *toska* does not mean to be in actual physical pain; neither is it isomorphic to depression. *Toska* is an aesthetic and conceptual experience of the world, a type of sight, a form of nondisinterested contemplation. Anguish is nondiminutive; it does not paralyze by isolating into an individual unit of selfhood and resists localization and instrumentalization simply into feelings. Rather, anguish is a collective experience of environments, made of layers, groupings and processes, plants and paperwork, splinters and icicles. Its individuating capacity is not working toward the subjectivation of the subjects of modernity but toward further differentiation in relation to the actual conditions of life and future unfoldings of the changing virtual. Both slow and silent reconfigurations of life and the rapid unfolding of moments of the unknown can change the consistency of the virtual, altering the future.[30] If devastation refers to the desertification of the virtual, in other operations it is frayed. In this sense of strands becoming unwoven and attenuated, anguish can also be undergone when the possibilities of the future suddenly expand.

Anguish, appropriated at the threshold of individuation, can thus be affirmative in a very Nietzschean sense—exuberant with a special variety of power, one that can be perilous and radical, or entirely nonpowerful. This force may—nonpowerfully—render something more powerful by creating conditions for the birth of percepts, actions, and happenings, however bleak or perilous. There is a richness of boundlessness lying in anguish.

An ecological and individuating understanding of anguish is exemplified in its outcropping in something known as a heroic act: another twist in the continuum of anguish. In the war against fascism, more than four hundred Soviet soldiers performed heroic acts similar to that of Alexander Matrosov, aged nineteen, who in 1943 threw himself at a machine-gun embrasure, blocking it with his body, thus saving his comrades and letting them complete the attack. In post-Soviet times, doubts were raised as to whether the act was accidental, whereby the wounded Matrosov fell onto the hatch of the embrasure in a way that rendered the machine-gunner incapable of removing his body. Multiple instances of similar events are often made publicly manifest when a political value can be extracted from them by inflation or by such doubt; another example is a recent local Russian newsmaker, an eighteen-year-old girl who disarmed and seized a villain trying to rob the accrued fares from the driver of an intercity coach. The young woman did not have any fighting skills and was the only person in a bus packed full of passengers to act in this situation. She later said that she did not know why or how she engaged in a fight and that she could not recall it either.

We consider these to be examples of a situation in which the force of a person's individuation, and its nesting in wider ecologies of force, brings the subject to the verge of annihilation, or the threat of it. Like some fetuses in pregnancy that poison or kill the mother's body against their best interests, a person or thing may have to become someone or something they cannot contain, something larger than themselves, bringing a possible arrest to their own life. In such a situation, the "have to" is not of the order of fate, but neither is it solely something that can be decided, so it gets brushed away as heroism, passion, accident, dysfunction, or fault but may possibly be better rendered through a cartography of luck, the theme of a later chapter.

Here it is important to grasp anguish not as a quality that is experienced as a direct effect of something happening—a feeling, consequently registered in the organism and traced in laboratories—but as a forceful experience that traverses and seizes human and nonhuman alike. Anguish involves a kind of individuation that can be obliterating for those through which it unfolds but is not necessarily so. Such experience can be triggered, as in devastations such as the spill of oil from BP's Deepwater Horizon. But they are often not framable as events, nor as the result of the choice of those affected or those who inflict them, and their spatial and temporal contours and time definitions remain uncertain. Neither the framework of heroism as a horizon for action nor that of the accident as a cause offer sufficient grounds for understanding these processes and yet they keep happening, sometimes gilded by irresolvability.

An example in line with the above from the realm of animals and insects can be drawn from something called adaptive suicide. Scott Forbes tackles this topic in his entertaining *A Natural History of Families*, an inquiry into infanticide, siblicide, and other such things as effective evolutionary mechanisms. Forbes focuses in particular on pea aphids that produce genetically identical nymph that are vulnerable to predators (such as ladybirds) or parasites (such as braconid wasps that inject eggs into living victims).[31] Upon hatching, the braconid wasp larva devours its host from the inside and, upon reaching maturity, bursts forth from the carcass to begin a search for new aphids to parasite. It chooses those nearby, likely genetically identical copies of the prior victim. Therefore, when one aphid becomes parasitized, it contains the seeds of the destruction of all of its genetically closest relatives. What happens then is that sensing aphid alarm pheromone or when approached by certain predators, a parasitized aphid will leap off the plant (though not immediately upon just being parasitized, as it can still produce some genetically intact offspring for a while; and older aphids do it more rarely than younger ones).[32] The response of the aphid—to leap from the plant—is a suicide, rare in the natural world, and made rarer by the fact that pea aphids are not social insects. The aleatory unfolding of an aphid life is overcome here by the involvement of an evolutionary adaptation, which is called upon to provide a clearer form of causation. Still, such extreme suicidal behavior in insects produces a drama outside of drama, scaling death between the individual, its genetic milieu, and the population. The seeming naturalness of such movements across scales, of nondramatic dramatic encounters is something that Muratova often brings to light.

Some of Muratova's themes are cruelty and meaninglessness, solitude and the inability to communicate while speaking face to face, but there is always something extra in her work. Muratova is a master of anguish. Her anguish is shared between humans, nonhumans, things rendered dead and alive randomly even when such rendering is orchestrated. Here, anguish has an architecture, its own mechanisms. In Muratova's films, anguish moves across scales and entities, making them do or not do things: individuating actions. Anguish is not centered enough to represent the problem or generate a solution; it is eclectic, mutant, engorging, and collapsing. In the last segment of Muratova's *Three Stories*, the girl is never called by her name, and there is no dialogue, engagement, or action but a series of semidirected activities: the repeated taking off and putting on of her dress, hitting a doll or playing with it, sneezing and yawning, getting angry, having arguments with herself. There is a half-expressed social conflict: the girl's mother, as the girl repeats unthinkingly, is expecting to be allocated the old man's room after he dies, and the man only looks after the girl while her mother is at work in exchange for her bringing him food since he cannot leave home. The man is also scornful of the girl's poor preschool educational level and of the fact that her mother takes food back home from the canteen where she works, effectively stealing it. The adults are of different sociocultural formations, and yet there is no properly expressed or ripe sociopolitical confrontation. The meanings of the episode are half-torn, ragged. It is one of the cuts of entanglement in which living matter immerses itself. Misplaced social conflicts are lived through the body of a disabled old man and a not yet enabled little girl. These conflicts are of matter, as well as of culture, enacted by one person lingering at the exit and another before the entry into the cultural realm.

Old age and youth are certainly among Muratova's interests—not because they are boundary states but due to their normality in a non-goal-oriented understanding of living. If proportionally, old and young ages take such a large part of life, their reason surely is not to lead to and from the period of responsibility, health, and a sane mind, if such are ever experienced, but to exist in their own right and purpose, as occasions of complex living matter. The struggles of such different times and conditions outside of morality, on the boundary of ethics, outside of reason, and across animals and objects are also conditions of anguish.

Muratova's anguish has a naturalness to it. Things "just" occur and yet have consistency to them as a common quality of matter living today. Core to Muratova's formation of anguish is the nonripeness, greenness, nondecisiveness of such occurrences. These are occurrences that kill, and yet they proceed without thought, reason, or much effort. Truly beyond good and evil, this naturalness is about the flow of time and qualities of interactions, some of which turn out to be terminal for those involved: chickens, humans, rats, apples.

Anguish here is different from that described in a heroic act; it appears more purposeless, less prone to resolution (even if it is annihilatory). As an experiential sensation, individuating and capable of generating action, it is drawn out, changing intensity. Yet different inhabitation of time and concentration of intensity, whether due to decision-making capacity, affective capture, genetic mechanisms, or other factors (such as the erasure of the fungi that might digest dead wood in Chernobyl, as discussed in the previous chapter), attest to anguish as an experience fitting the complexity of today's vociferous and stealthy societal and ecological unfolding.

Mikhail Gasparov, an outstanding philologist and literary scholar, once said something along the lines of the following: "I don't feel that I have human rights, apart from the human right to starve."[33] It is this sense of the normality and indeed centrality of something often "hopefully" called a bad patch in life, or better, a run of bad luck, that finds itself at odds with the expectations of the current First World human. In Gasparov's phrase, starving, dying in a ditch, is not atrocious. Though often preventable by measures such as social security or collective health care, more equality and less racism, something fought for with socialist and other ideas, the statement is not evidence of a shattering of values. What Gasparov does is to deny himself a privileged position in an ecology of anguish, a position that would take him out of how others might undergo their death, disappearance, or irreversible change.

To starve is a function that is shared with other animals or plants. To die in a *ditch* is already to have a relation to the human, and thus, in this case, to an abnegation of justice. To starve as a human in a condition of plenty is equivalent to dying in a ditch. To starve as an animal in a condition of climate damage marks the gutting of the world. In an interview, Muratova said that to love or hate people, one must occupy a position of being either above or below them, and that she is simply one among them, just as Gasparov effectively situates himself among animals and plants.[34] Such nonnormative amongness rather than estrangement is also the condition of anguish.

It is a widespread understanding that there is an affinity of *toska* with a yearning for the unattainable (a companion to the unspeakable addressed earlier). The aesthetic qualities of *toska* align it with some kinds of thriving in art. But in *Parting from Phantoms,* Christa Wolf writes, paraphrasing Heinrich Böll, "Art is inconsolable . . . which is not the same as desolate."[35] Anguish is not the yearning for the unattainable, for some kind of metaphysical surplus not marred by earthly scrabble; it is the sniffing out of the inconsolable. Being unable to be consoled is not necessarily a willful refusal, a caprice or disappointment. Consolation may not be available because teleology could not be located, because the accelerated unfolding of events does not provide space and time for it, or because there is no outside when one is among and within.

Multiple entities are anguished differently as their unique and changing compositions unfold in space and time, not quite moving in any "good," or single, or clearly definable direction. If nonconsolation is a feature of anguish, being among things is its method. The experience of blockages or alterations to the divulgement of life, through moments or dragged-out processing of futures, may have numerous ethico-aesthetic consequences, only one of which is anguish. Related symptoms were earlier appropriated as black or yellow bile, melancholy, *toska*. Framed as depression, and reduced to a matter of medical management, anguish lurks at the level of the mind, emotions, hormones, promises. Toxic ecologies, suffering living beings, mutant thriving forms of life, or entities without a life span may also experience a recomposition of anguish.

PURE IMMANENCE

Muratova's unsentimental depiction of life has affinities with Deleuze's figuration of life as *Pure Immanence*. In the essay of that name, Deleuze writes about the subject and object being simultaneously born in the processes of actualization, and it is then that they obtain their transcendent positions, positions of "denaturalization," of "being locked." The subjective and objective "falling outside the plane of immanence," as "accidents of internal and external life," are kinds of occurrences that are not even necessary: life, in his vocabulary, can happen without individualities, just simply with singularities.[36] Muratova tends to create instances of the singular, often without or half-outside individuality. She operates on that border of formation or decay of individualities: on the level of accidents, occurrences of immanence.

But what is anguish in relation to vitalist pure immanence? Is it related to the occurrence of becoming locked into a subject and an object? Is it only about an episode or an event of becoming pushed into another mode of existence, out of a subjectivity—a radical or slow change of a particular singularity's range of trajectories? Deleuze says that pure immanence, the life that small children are imbued with, is pure power and bliss. The crux of this chapter is to offer a figure of the way in which the blissful power of immanence that is always in a process of singularization, of actualization, can be anguishing.

A principle of affirmation, according to Deleuze, consists in the following: "Immanence is opposed to any eminence of the cause, any negative theology, any method of analogy, any hierarchical conception of the world. With immanence all is affirmation."[37] When Eugene Thacker tracks the logical unfolding of the ontology of immanence and univocity through Thomas Aquinas, Johannes Scotus Eriugena, John Duns Scotus, and Nicholas de Cusa to Spinoza and Deleuze in his book *After Life*, he proposes a form of "dark pantheism" as a way of thinking its logical challenges. Thacker argues that for life to be univocal and immanent, it has to be generous (productive, pouring forth), conditioning (life is irreducible and conditions its instances while being fully immanent in them), pervasive (life is everywhere and not in itself in any single manifestation), and expressive (a causality of univocity, in which "a Creator is fully immanent in all creatures").[38] In Deleuzian ontology, constant creation, or immanence, is affirmative.[39] Affirmation is not opposed to negation and not neutral but is intensive differentiation itself—always creating, generative, superlative—and joyful, in a Nietzschean way.

Affirmation is what allows Deleuze to assign "higher qualities" to life, to have the virtual, which is real and immanent and yet can act as a source of life. Here life is generosity, "an excess" of being, larger than its own being.[40] Thacker shows that generosity, excess of expression, of creation, is the ontological condition upon which life can coincide with the principle of life for a fully immanent system. Hence, affirmation is an ontological principle that solves a philosophical conundrum.[41]

The challenges of such a univocal system of coordinates, according to Thacker, are radical flatness (where famously a tick is potentially equal to God), making the establishment of "relations between the orders of being" difficult; the problem of invention; and the unavailability of negation. All of those are reactions that have been voiced against pantheist philosophical discourses over time.[42] One of the critiques or consequences of pantheism that Thacker tracks and one of the reasons why he calls it "dark" is its unhumanness. In this, there is a relation to the earlier themes of deep ecology, but work on them is done via different routes. The radical non-anthropomorphism of this neutrality can also be read as misanthropy. Muratova's engagement with the nonhuman and neutral processes of life seem to be aesthetic manifestations of these principles: "Pantheism in Deleuze's sense points to a horizon in which both 'life' and 'thought' can be understood in non-anthropomorphic ways. . . . The divine is understood to be indissociable from nature, and because of this, radically unhuman, anonymous and neutral. This pantheism is certainly far from the pantheon of Greek divinities."[43] In the second segment of Muratova's *Three Stories*, the lead actress says, "This planet would get a grade of *nil*."

However, such neutrality need not only apply to humans. In such affirmation, suffering, pain, meaninglessness, and obliteration afflict animals and nonanimals alike. Anguish is a modality of experience, among life, of the alterations of the course of life in its principle of forever overflowing inexhaustibility. As we have proposed, anguish has a genealogy and changes over time. The changes in vitality pouring forth are part of what distinguishes contemporary anguish.

Thacker suggests that the modality of affirmation presupposes within itself that of negation, nihil. Affirmation of contradiction involves the movement from all to nothing within the all, or both at once—all and nothing: "What enables the plenitude is precisely the void."[44] Thinking nihil, the void then for him is embedded within immanence of affirmation.

But being among means the void is not available, that devastation is not complete. Even with a *nil,* the planet is still there, annoyingly. Anguish itself is not negation, nor nihil. If it is unhuman, it is also unanimal and unthing. After all, the bliss of affirmation, its jouissance is a kind of enjoyment, which in its plays with the virtual can be annihilatory or harsh, and may be just another word for anguish. That is why anguish is a bleak joy.

ANGUISH OF APPARATUS

Perhaps anguish can be seen as something empathic that can be mutually shared, or, like those of other ambiguous sorts, exist as a commons. Here, it is the condition of amongness that requires inhabitation. Maybe we need to learn to be capable of being in anguish in rich ways and at different levels in order to learn to respond to today's world adequately, without withdrawing into mere unhappy negation, ignorance, or oblivion. Instead of an economy of happiness there are alternative economies of anguish, ones that are not correspondent to emotional descriptors and medical prescriptions and do not sooth themselves by calling upon the apocalypse. One could also say that what is called creativity in cognitive capitalism does not make people happy. People have enormous reservoirs of potential, not only for being unhappy but being outside the project of happiness.

There is a classic argument to be had, in a way, around modes of resistance that consist of diseases and deaths. If disciplinary regimes were challenged by hysteria, madness, and prostitution, then biopolitical regimes of collapsing ecological conditions are affronted by various cancers and complex new immunological diseases, where children's cancers are just too vile to be regarded as an ecological answer to global failures. Here, an increase in cancer is one of the material responses of autodestructive resistances to the proliferation of chemical or other hazards. It is cruel politics—to see a dying child's body as a protest, a witness against chemical warfare and ecological collapse. The inferno of such occasions resists interpretation and appropriation but is part of the churning of today's capitalist, ecological, financial, chemical, industrial, and informational apparatuses, yet such accidents are rendered inconsequential and irresolvable.

Anguish traverses scales, inhabiting an ensemble of elements, objects, and forces. The little girl's preparation of the poison in *Three Stories* is a chance assemblage and interaction of elements, facts, and processes: things align together to experience anguish. Anguish is an ecological condition, becoming manifest in habitats and systems. It can also be said to have the potential to acquire systematicity itself. There are apparatuses that become filled with anguish, such as a hospital becoming infested with bacteria. One would very much doubt the presupposition of a possibility of the existence of health, but the specificities of the contemporarily dominant forms of governance, organization, and mobilization certainly amplify and create new modalities of anguish. Such amplification does not mean that anguish today becomes more intense or more total but that it complexifies by breaking scales and making points of connection that are nonlinear, irregular, and of different orders.

A clear illustration here is *65 Red Roses,* a blog by Eva Markvoort from Canada that she maintained from 2006 until 2010, when she died at age twenty-five.[45] The blog was dedicated to her life with cystic fibrosis and helped raise awareness about the disease. An archive of her feelings and thoughts, treatments and actions, this is also an archive of her body, the assemblage of its various parts, such as the eyes and lungs, and of her suffering. This blog is not only a documentation of the feelings of a dying person, nor a memoir or a letter to posterity. Here, Eva's body is the sufferer and the witness—related to the humanitarian witnessing that becomes a mere document as opposed to testimony that is emotional. This blog, which is an annals of anguish and tragedy, is also witness of the anguish of her organs, lungs, and respiratory system. They are present in scans, X-rays, and videos, acquiring an agency of their own. All these anguished entities and the organisms they compose are locked into, translated, and emerge as networks that extend life in semiotic, algorithmic, visual forms beyond the dead body. Such blogs and other records of ill or dead people travel as anguish enlists communicational, mediatic, and computational forms into its assemblages. The biological, ecological, evolutionary, and industrial structures that bring about these deaths also become part of the apparatuses of anguish. But such scaling is not uniform or smooth; anguish is not a flow. And perhaps our imperative should be to invent ecological, ethico-aesthetic forms that do not flow and that would be adequate to modalities of anguish that we are immersed in but refuse to see.

FROM *TOSKA* TO PLANETARY ANGUISH

Deleuze observes that the Cartesian idiot (the thinker of idiomatic thoughts) went crazy in nineteenth-century Russia, becoming creative, a craftsperson. Between the twentieth and twenty-first centuries, *toska* mutated into anguish, as it became a global—both distributed and nonhuman—experience of ecological disaster and other changes to the planet. Perhaps an aspect of *toska* has always been about the changing availability of another future and another now? As large-scale, complex, cross-species, nonlinear, clandestine arrangements of the living and the nonliving experience and undergo reconfiguration of the future, the withdrawal of another soon, or negation of proximal development, anguish becomes a basic and nutritious state and sensation, an ethico-aesthetic sensibility commensurate with today.

The figure of *toska* was initially born in relation to the question of humaneness. The philologist and cultural historian Sergei Averintsev said: "The twentieth century compromised the answers but didn't solve the questions," and such questions (of social and economic inequality, discrimination and exploitation) are further expanded and radicalized today to include climate damage, the plenitude of affirmation of plastics, spills, and the unequal commons of toxic waste.[46] The answers of Nietzsche—in relation to the human—were a defiance of the guilt and fault mechanisms of the moral order, and the generation of another form of value. Deleuze's answers were—in relation to the cosmos—an ontological order of affirmation. Both of them advocated production and generation: of other values, other causes, other change.

Toska was embedded in the literature of Fyodor Dostoevsky and Anton Chekhov—and was drawn upon in the late decadence and early modernism of writers such as Henrik Ibsen and George Bernard Shaw. Anguish, on the other hand, is displaced, marginalized, outsourced, nondescribable, unregistrable. It is molded into the apparition of a failure in meaning, causality, and agency. If in modernity even melancholy becomes cheerful as its only way of being relevant, then anguish is muffled by the hysterical cheerfulness of cognitive capitalism in the midst of ecological attrition. Anguish, meanwhile, is not opposed to happiness; it just has nothing to do with either happiness or unhappiness.

There is a lot of striving for anguish today, yet few deal with it or rely on its force. Ours is a period determined to be soothed, whereas what we need is more recognition of anguish. The embrace of anguish also means an affirmation of the possibility of production: anguish may accompany attrition but also expansion.

In anguish, futures can also be multiplied. Political projects may spring up on the basis of anguish. They would need to be nonlinear, multiscalar, coalescent, opaque, disjunctured, embedded, unknowable, slow and rapid, and very lucky.

Irresolvability

An alternate history of literature can be traced through narratives concerning problems that face no likely prospect of the notion of choice ever being of a wise or healthy kind. The novel is a place where uneasy, furtive consciousnesses come into bloom: Ivan Goncharov's hesitating *Oblomov*; Fyodor Dostoyevsky's anguished souls; Malcolm Lowry and the evasions of booze; Samuel Beckett's labyrinths of impossibilities; the multiscalar fracture lines that make up the jagged sentences, plots, and persons of Elfriede Jelinek; the immense sprawling sentences that struggle to draw the world of László Krasznahorkai; Rachel Cusk's acid diagnostics; Svetlana Alexievich's atlases of collapse. The novel at its peak as a media system is one of the interior folded into the outside: "Thinking with someone else's brain. Schopenhauer called reading," recounts David Markson.[1] To use someone else's brain is a relief, like a day spent staying in a hotel where all your familiar shit is absent and where what is there is at least well ordered, even if, at best, slightly suspect. But in the text Markson riffs off, Schopenhauer was thinking of the difficulty of establishing systematic thought in the conditions of a stream of literary babel.[2]

Christa Wolf, to borrow her brain, has pertinent things to say about another narrative structure, the Cold War.

> What made stone-age people or primitive farmers unhappy was different in kind from the misfortunes of modern men and women. There is

no way they could have felt the terrible demands of conscience we feel when we see that we cannot avoid making a decision but that none of our choices is the right one.[3]

Here, she poses the condition of choice on an earth lit unbearably clearly, not with the optimism of Hegel's spirit but by a now radioactive "sunburst, which, in one flash, illuminates the features of the new world."[4] In this flesh-charring light, the state of irresolvability is that of a conflict mapped out by other peoples' brains, other larval mentalities, and, among them, the then emerging electronic ones.[5]

VOID UTILITY

The condition of irresolvability—which arises when, as Wolf says, *we cannot avoid making a decision but that none of our choices is the right one*— is worth looking at because there is something about the present that seems to generalize this condition rather than solely act it out at the scale of states, of weapons systems, and of strategy, or indeed at the level of choice per se. In such a condition, choice is hypostasized as actually possible. Whether choice is actually conceivable as rational, bounded, or linked to a conscious subject is left as an implementation detail. Indeed, choice as a transcendental that persists despite the lack of agents able to exemplify or to address it is a fundamental characteristic of political theories that attempt to establish its freedom, where freedom of choice means the increased autonomy of choice from any agent capable of making it. Interwoven with such a condition, there is an increased intractability of the world for those who ostensibly, as humans, comply with the definition of sensemaking agents. This chapter argues that irresolvability has become a fundamental means of structuring life in the present. Such a consistency of formation of the world in which things cannot be unified but are sustained as systemically irresolvable is a mode of operation of power that calculates devastations and besieges anguish unless it can be usefully employed in the maintenance of nondecidability. One of the ways into the question of irresolvability is by working through some of the technical, philosophical, and cultural conditions and precursors of the misfortune that Wolf wrote about. There is also the question of how this condition developed further from the time in which she wrote.

In some recent cultural theory and philosophy there is a turn to the discussion of finitude, extinction, the void, and the negative (as partially discussed in the previous chapters). Uneasiness is turned into texts, a means of establishing a relationship to disaster by turning it into something that is properly and, once the correct philosophical steps have been taken, evidentially factorized as being unknowable. Certain in its difficulty of submitting to knowledge, the void, for instance, to name one such entity, is calculable in its omnipresence. We suspect though that this is all probably too meaningful, too decisive, to fully account for the lushly variegated crapness of the present. One of the cultural reasons for the attention to the void is perhaps an articulation of the wider sense of the amazing stupidity of that part of the human race with any capacity to resolve the situation, in their inaction on the question of climate damage. In this condition there is the inversion of what has become a theoretical commonplace. It is more complex and more intellectually satisfying to talk about the obliteration of everything than to state the now entirely obvious position that the chemical outfall of the contemporarily predominant fuels and diets, and the economic, social, and political forms that defend and subtend them, is tending toward the suicidal, something that ought to be acknowledged if not actually addressed.[6]

The void makes too much of a proper fulcrum to turn thought on. What is needed is something less dependable than nothingness, which after all is virtually a utility, readily on tap. The void has become a destination like a Friday club night, overexcitably cluttered with academic outputting. It is cheerful, popular, if a bit fetid, and, given the hope for a final obliteration poured into it, never entirely exhausted in its attractions: though, perhaps there is room for a more minor mood of preemptive indeterminacy. It may be then that we are arguing that irresolvability by contrast is merely dismal, more authentically of genuinely poor quality, rather than something too easily ponderous and dramatic, and for that we apologize.

INFRASTRUCTURES OF FEELING

What we want to discuss in relation to irresolvability is something that moves across the scales of subjects and states. Irresolvability is the structural incapacity to fix a problem. It is a foundational uneasiness, one induced by a problem that converges in your being but is way beyond it.

Further, irresolvability is, we argue, a policy of generalizing the economy of deterrence into a presiding modus operandi of the present. Irresolvability names the condition in which the structuring incapacity of action of the Cold War becomes—by means of related technologies, economic and organizational forms, and processes of subjectivation—a part of everyday infrastructure of feeling. If *structures of feeling* in the work of Raymond Williams are to do with both what is possible to think and feel in any given time and its incorporation, as a form of hegemony, in the locale, and the activations of place as culture, *infrastructures of feeling* are established by medial and strategic forms that integrate spaces and times as conjoint conditions of possibility.[7] The Cold War is one of these; climate damage is another.

The novelist Christa Wolf was first a teenager in the Third Reich, then a communist, and thereafter a renowned dissident in the German Democratic Republic. Stunned by rampant capitalism, the time of "reunification" also revealed both that the Stasi had not only spied on her, generating boxes of dossier material, but had also considered itself to have recruited her as an informal collaborator (the evidence for this is one thin file), of which she had no recollection to offer. Forced into making a homeland in her books in the last decade of her life, she wrote about the condition of irresolvability while reflecting on a period of time spent in the United States, in Los Angeles, a city whose sprawling condition, driven by the logic of the intersections of certain machines—capital, cars, the realized dream of sunshine, possessive individualism, the openness of the desert, the mechanics of celebrity, racial demarcation, and so on—perhaps echoed some of the amorphous state she notes.[8] Wolf writes of a condition in which no decision can be right. Here, the world is composed as a generalization of wrong choices. Perhaps what is novel about the tendencies of the present is that even the capacity to fully make a choice, even a wrong one, tends to be uneasy and, if not fully foreclosed, too vaguely multifactorial to determine: here the world needs a decision support structure.

As an analysis of these conditions, this chapter proposes that there is a genealogy to be drawn between the following: the game theory of the 1940s and 1950s and its relation to the Cold War and the state of intractability Wolf describes; the theories of self-organization of the 1960s and the great interest in them taken by Friedrich Hayek, among others, which is specifically articulated in his figure of the ideal market that he calls a *catallaxy*; and the development of irresolvability as an ostensibly emergent and multiscalar quality of contemporary life once those theories have been more or less ineptly operationalized, even if only ideationally, as players in economic and subjectival life. Such an account is not to be taken as the proposition of a historical fact but as the wager of a connection. Irresolvability is also erected in art and literature, as visually and spatially incohereable shearings of multiple surfaces, the unconscious proliferation of multidirectional debris, and inhabitation-in-advance of the combinatorics of all possibilities. This chapter therefore is of a systematization of a kind, related to that which Schopenhauer called for, but is also one that operates precisely by integrating babel rather than rising above it.

DETERRENCE COMES FROM ABOVE

The particular generalization Wolf writes about is that of the Cold War and the spate of mutual excitation that two geopolitical blocs brought themselves into via means including technology and novel forms of mathematical logic. This is an interesting condition to think in, since for writers such as Martin Heidegger, the nuclear condition effectively *defined* technology as a mentality.[9] For theorists of technology, reading Heidegger is of course to encounter the repulsive uncle whose kisses of greeting are both sufficiently crushing and solipsistic and moistly intrusive to make one feel that you haven't done *something* ill enough to perhaps deserve them, and maybe even like them under the wrong conditions.

In the "Letter on Humanism" that Heidegger frames as a response to Sartre but in which he rejects the question of choice in relation to the entanglements of being and essence, there is a differentiation of thought from technical thinking. Such thinking sets up the "encounter with beings" in a "calculative businesslike way," one that carries over into a philosophy increasingly fixated on "explanations and proofs."[10] Disdain for such things is articulated in Heidegger partly as a means of preserving an atavistic longing for an irrecuperable whole that dismantles the prospect of actually being able to engage with such conditions. Nevertheless, the letter points toward the incorporation of technocratic decision structures into wider forms of life.

Wolf regards the "misfortunes of modern men and women" as partially a side effect of the Cold War, whose mentalities were probed, conjured, and installed by numerous means of varying degrees of abstraction. It is this relation between modalities of abstract structuration (in strategic logics and in idealizations of socioeconomic forms) and worldly events ranging from the subjectival to the global or cosmic, and their consequent means of divining and inaugurating the possible and impossible, that we are after here.

To trace one site of the infrastructure of feeling of the Cold War, one can examine the Manhattan Project, an advanced example of the communization of intellect toward the generation of new kinds of weapons: the first atomic bombs. It was a space marked by the interrelation of highly novel and arcane mathematical and physical calculation coupled with the arrangement of materials, instruments, equipment, metals, of greater and lesser degrees of refinement; a space whose ligaments spread to the Shinkolobwe mine in Congo, and which were fed by fleeing Europeans and exuberant budgets; a space set in the sprawling landscapes of North American military bases, university campuses, and in those of atomic particles. This scalar exuberance is found too in the calculation of the activity of atoms in a state of great expansion, coupled with the basic violent gesture of dropping a lump of something heavy onto someone else.

What followed, in Hiroshima and Nagasaki, was a disaster. And what followed in terms of perpetual strategic foreplay and massification was also the attempt to map and articulate the consequences of such a disaster, to separate it as a condition of an irremovable possibility of indiscriminate annihilation, out of which there is no way and outside of which it is impossible to think. In this newly global infrastructure of feeling, some of the same mathematicians who were involved in the project came to the fore in marking out this terrain, using variations on work that had been generated beforehand.

Oskar Morgenstern and John von Neumann's *Theory of Games and Economic Behaviour* as a founding text of game theory is a magnificent attempt to stabilize events, to understand them, but also to make action on them tractable by means of mathematical logic.[11] It is an attempt to make reality yield by giving it instructions, by laying out its secret laws that until then it had never known it had, and to possess them, to seduce them so that the rigor of the secret paths of strategy become intimately known in a way that convinces themselves that they exist. The book itself became part of a much larger assemblage, of which it was something of a sublimated keystone never much revisited in von Neumann's own work, and which grew out to involve military technologies, research organizations, and what can be said to be a rhetorical and subjectival infrastructure of threat, opportunity, counterthreat, and a calculus, more generally, of a governance of choice.[12] This was a structure of feeling, localized within the pages of a book, that had the novelty of implying subjectivation with strategic importance but also of a governance that in turn implies subjects who make that choice and the ruling apparatus in which they do so.[13] It was also, we propose, a preliminary manual for the establishment of irresolvability in disturbed conditions, a device to register the world in terms of both transparency and information and to orient in the state of their unachievability. The book emphasizes strategic thinking as a basic principle of interaction between entities in the world, one based on asymmetric competition and collaboration with absolute conditions of winning and losing—zero-sum games—being highly emphasized, even if only as a fantasy that keeps the game in motion. A little later other formations, such as the celebrated Nash Equilibrium, also came into circulation as models of metastability in which no absolute victory need take place.[14] The contours of irresolvability in the Cold War are sketched in these documents and developed further in texts such as Thomas Schelling's *The Strategy of Conflict* that in chipper and concise language mapped the terrain of deterrence, a means of positively affirming irresolvability as an end.[15]

Game theory was only one of the mechanisms for choice developed during the Cold War period, in which a multitude of agendas, opportunisms, interesting leads, and seams of possibility for the elaboration of a certain modality of superiority were mobilized in order to explore the phase space of power at that historical moment. This was a historical moment in which mathematical, electronic, logistical, and other formulations came into being in competition and interdisciplinary interaction with each other; saturated in cash from military largesse, such competition was displaced from being merely economic. Given this state of abundance, there was a necessity of mobilization. Others were to be enrolled in strategies of conflict and become part of the strategic wager: these were members of populations who are arranged as targets or as actors whose processing of the situation correspond to some model of interest, some formulae of rationality or nonrationality. One way of handling irresolvability is to enroll more actors, other factors, to spread the circuit and the distribution of probabilities but also to extend these across the surface of the planet and through time. At that moment, across and beneath that surface, the Krasnogorskiy mine in Kazakhstan starts to disgorge its uranium in complement to sources of the substance commanded by those who describe themselves as leaders of the curiously named free world.

At the same time, this process further rends the predictability of monopolies. History comes to a halt if or when the atomic weapons are used, but it also starts to feed into and corral the passages of time.[16] In turn, other actors—Britain, France, China, India, Pakistan, Israel, North Korea, Iran, and the regional proxies of the countries affecting to project themselves as superpowers and thus accede to nuclear "protection"—decenter and fractalize the generalized game of "Chicken," as Bertrand Russell described the arms race.[17] This game is in turn rendered more quotidian, factored into models, and becomes a new norm against which further acts of chicken, or brinksmanship, can be played out. The Non-Proliferation Treaty, the International Atomic Energy Authority, and their apparatuses create further conditions in which things stabilize or are rendered as appearing to do so. The fantasmagoria of weapons of mass destruction in Iraq culminated in their use as a pretext for invasion in 2003. At the same time, technical game changers or disruptive technologies—such as the introduction of the hydrogen bomb, intercontinental ballistic missiles, and "Star Wars"—attempt to shift or displace the axes of tension. Deterrence then is how those who deploy the strategy explain it to themselves as they generate all this infrastructure, and irresolvability is how it is experienced by those they enroll.

CHOICE STRUCTURES

Christa Wolf's figuration of the generalization of wrong choices, one of which must necessarily be taken, has potential affinities with some of the central questions of Jean-Paul Sartre and Søren Kierkegaard. But Wolf nods toward a condition that diverges from the one they map—in that choice is now often inhabited by cybernetic operators, choice structures that provide an architecture of decision that protects against the anxiety of being. This change is a distinct shift. Choice moves from being the result of tradition, underwritten by belief systems or customs such as religion, to being in the modern world, an awful imperative to confront one's own fate where there is no longer any moral armature. This emptiness, which is at the same time a certain image of freedom, is subsequently populated by logical models that provide convenient rails or recommendations for thoughts to run along.

Paradigmatic of a certain strand of modernism, Sartre's conception of "anguish," distinctly different from that described in the previous chapter, is based around an impossible choice, out of which a decision must be taken.[18] Sartre's is a conception of the subject that self-generates out of manifold forms of will and that is fundamentally engrossed in its own subjectivity as it comes into composition with and is scarified by existing mechanisms of subjectivation, such as class formations. The existential dimension of the "Attunement" section of Kierkegaard's *Fear and Trembling* that Sartre draws on—in which appears the narrative of Abraham being asked to sacrifice his son to God, an obligation he is relieved from only at the last moment—is excruciatingly harsh. The text is notable as a piece of philosophical writing in that it presents a filmic cut, from one point of view to another, in an unusually pared down and vivid manner. Kierkegaard's "dread" resonates here with Sartre's anguish. The dreadful choice here is to separate from the child, sundering one linked organism from another by annihilating it. A decision in this calculus must solely be rational, and reason under such a schema negates feelings such as dread.

Sartre poses the individual choice as one that at the same time proposes itself as a template for all of humanity: it is not that the same choice must be made but the proposed urgency and ethos of it should mark the mode of decision. This dual state of choice, which is both individual and universal and hence laden with responsibility in a condition where everything is permitted, rends the anguished person but is also their condition of action. It is a condition that may be immensely difficult to inhabit but is also one of new figurations of becoming. In Sartre's immense biography of Gustave Flaubert, for instance, it is Flaubert's struggle against himself as a bourgeois subject that sets up the means of escape through art and that articulates in turn, as a variant potential "universal," the modern figure of the artist.[19] Under this schema, since a choice must be made, irresolvability cannot exist.

Sartre's Flaubert and Sartre are lucky fellows. They are modern, in that there is a reflexivity in their formulation as subjects that extends beyond their corporeal being, but they are also heroic in the classic sense, in which they "must" do something and where action is possible, however potentially disastrous it might be. This is a crucial part of the architectonics of sensibility in the modern period, where the disenchantment of the world in all its confusion and groundlessness must be answered by a subject that determines its own course. Turning decisions into the operation of logical automata is one way out of this condition, but it has its precursors. Rich seams of paradox existed in the classical world in terms of logic (for instance, those of Zeno), but paradox never appeared in mythology—that always resolved ambiguity. What contributes to the sensibility of the present world is that now such paradoxes, in weaponized form, exist not only as our fundamental myths but as the realities those myths distill. The condition of finding oneself raised under the rule of two contradictory but inevitable imperatives—*you must make the right decision; none of the decisions you can make is the right one*—is typical of a condition that, for Gregory Bateson, produces schizophrenia: the double bind, a form of paradox. Bateson sets his argument out in relation to the condition of the family, specifically a configuration of the mother and child relation. But it has a more general applicability, and he later provides evidence for its cruel experimental inducement in dolphins, which *pace* Flaubert may, in Bateson's words, "promote creativity," when this is encoded as novel behaviors, such as those a starved and confused dolphin may untypically make when trying to work out what will induce a previously generous trainer to stop depriving it of food.[20] In this model, schizophrenia is incurred when a being is in a double bind yet denied the metacommunicative layer, the ability to communicate about communication, to theorize. The problem with being on the receiving end of deterrence is that it implies being forced to rationalize under conditions of absolute fear of nuclear annihilation as the grounding basis of reason; the metacommunicative layer is there and measured by the kiloton. Game theory is rooted in part as an attempt to abstract out from this layer into one of a strategic universality.

BABEL OF VECTORS

In contrast to the existential dimension, game theory occupies the timeless time of logic where decisions and calculations are not even immediate but instantaneous to the posing of the problem that they are coeval with.

More recently, models drawn from cognitive science, neuroscience, and affect theory cluster around the possibility that such an instantaneous time might also be operative in different ways within the nervous system.[21] Here, the subject is tangled up with the conscious sense of acting that often covers up or acts in tandem with a certain kind of automaticity. Since nervous systems have assessed and sorted an action before it becomes consciously registered, history happens before it is aware that it happens.[22]

In contrast to the immediate time of logic or of affect, there is a recognizable wave of the spatialization of irresolvability in responding to the conditions Wolf writes about, and one that is of a different order to the existential subject discussed above. Much of the Pop Art of the Cold War period is both diagnosis and symptom of this position, in turn gleefully neglecting any attention to the previously dominant essentialism in painting posed by movements such as Abstract Expressionism. In such work, the conditions of irresolvability become tokens, existential conditions, or iconic figures whose registration passes into the unconscious. For instance, the image of a simple map of the range of impact of a one-megaton nuclear bomb landing on London is marked, in detail, onto the central part of the surface of Derek Boshier's 1961 painting *England's Glory*.[23] The painting consists of repetitions of the surface of a familiar matchbox, the sheering calls to order of the British union flag and the American stars and stripes, dabs of paint, cartoons of clouds, stripes from a regimental tie or a dodgy bit of patterning, faded gray on white window frames in a detail within a swag of paint, framing in turn the picture of a young woman. Nothing holds fast and what does cohere is vacuous at best.

Irresolvability, more broadly, becomes a condition that is registered in such works of Pop Art as a problem so deep that it is only traceable in overlapping surfaces. Veils, swags, and torsions of visual phrases collide but never fully cohere. The space race; dance crazes; consumerism; film monsters; pop heroes; advertising; the new convenience of toothpaste that can be squeezed glistening, white, and hygienic from its own plastic intestine—in short, all the heavy machinery of the spectacle come together with intense excitement, vacuousness, sensuality, hackneyed visual forms, new iconographies, vernaculars, processes of Americanization (especially of the United States) and ephemera, each making the condition of the other unstable, questionable in terms of what category it belongs to.[24] What is notable in such paintings is that they are all conjoined in ways that do not resolve as the surface of a single visual plane. Pop Art, it can be said, catalyzes from the twin aesthetic condition of irresolvability and ostensive abundance specifically crystallized in the 1960s.

This articulation of the incoherent image of multiple surfaces comes to the fore contemporarily in the work of Hito Steyerl, where clusters and swarms of discrete iconographic and representational forms infest the screen; Ryan Trecartin, where iconographic, verbal, and personological monstrousness stagger across it; and Ed Atkins, where the filmy, cluttering, and glutinous layers generated by effects programs slide over each other in a relentless state of abundance. Such work, also like that of JODI or Joan Leandre, works through and brings together computational processes and platforms, graphic vernaculars, and ready-mades generated and modulated by software. Irresolvability augments semiotic shearing by drawing upon computational processes that collide in a visual plane as much as they do in the act of viewing.

Where Sartre delegates the labor of maintaining a cogito to the existential subject, other conditions scatter that place of work. One way of understanding the extent of the proliferation of stabilization efforts that such conditions entail is to follow them through as part of the consequences of the work of Kurt Goedel.[25] A full description of a system or a language cannot be completed within that language; such a thing thus cannot be controlled in and of itself. New languages and systematizations have to be layered and overlapped in concatenations and patchworks. It is this condition that also later marks some of the break between structuralism and poststructuralism and both the proliferation and the malaise of formalisms. What is remarkable about the "achievement" of the arms race then is that its proliferation is of one kind of system, and yet one that manages to provide a variously backgrounded or foregrounded total narrative.

GHOST LOGICS

Philip Mirowski assembles a history of economic modeling, policy, and their multifarious genealogy in *Machine Dreams: Economics Becomes a Cyborg Science*, a virtuoso mapping of the transition of logical formalisms gestated in game theory and related domains of mathematical description and formalization into active economic modeling and organizing forces, where the "metaphor of the computer" is "projected" onto economic conditions and modes of knowledge.[26] This projection is not, of course, coherent, full, or self-consistent but something that creates conditions of what, in Gilles Deleuze's reading of the work of Michel Foucault, is described as the deep "anonymous murmur" of society, and in which the self tendentially coincides with circulating "logics."[27] Such projections are not merely statements but actions. Instead of using linguistics as a model, such logics rely on the unfolding of material force in programs and on processing the projections of possibility. Projections of logics operate as structured anticipations of possibility but also as kinetic extension of force and articulation. Such anticipations are produced in formulae that address actual or possible actors, varying scales of possible actions, and the projections that those in turn entail and corroborate. In this arrangement, the functions of subject, object, and concept contain "derivatives."[28] These derivatives can of course be speculated upon, subject to interpretation. Ghosts may form among them in the mutual overlayering of projections, interpretations, and speculations upon the articulations of these. Such ghosts, and possibilities of them, gain traction on the present and preempt futures. Ghosts are a useful name for these murmuring entities, attractors in the shifting phase space of possibilities, since, in the mode of the Cold War, they are the accumulated traces of the projected dead, acting back onto the present. But they take other forms, become new objects that move as lugubrious statistical ectoplasm.

Irresolvability is not the result of work of a specific class of strategists unfolding a plan and a techne to have it implemented; neither is it simply a social order of a silent compulsion embedded in words and things.[29] It is a coming into being of arrangements that hinder, hamper, render difficult, or obscure the prospects of or the capacity to imagine a way beyond a particular impasse. They install a population of ghosts in the places where some of Sartre's generation saw the possibilities for the articulation of a decision.

Consider, in the twenty-first century, the voice of the European Central Bank, the International Monetary Fund, and the European Commission as it speaks to the young people of Europe or the voice of austerity as it speaks more widely: "Your life has been prelived by the money that came before you, the becoming of the negative imprint that you must endure through the duress of life as your debt." One of the forms and conditions in which irresolvability occurs is economic, where for instance it may, within a certain vocabulary, manifest as stagnation.[30] But irresolvability is also more crawlingly lively and dynamic: here it is useful to refer to Louise Amoore's and Brian Massumi's works on the formulation of preemptive governance. The logic of preemption is a decentralized form of deterrence discussed above: it takes the modality of the desire built into the arms race and extrapolates it into a proliferation of niches and microclimates.

Amoore valuably shows how techniques of accounting, the position of the consultant, and the calculation of possibility become integrated into an unfinished system of "proxy sovereigns" that unfold into and rig the deep border of contemporary life, "decid[ing] on thresholds of normality and deviance, limits of admissibility to citizenship and personhood."[31] The objects of such calculations, statistical outliers and their ghosts, the unpreforeseeable, come into various forms of manifestation, which are, in turn, only addressable via further calculations of possibility.

In his writing on preemption, Massumi traces the forms of experimentation certain states made after the cessation of the Cold War. Their logic is uncannily similar, hinging on the ontogenetic or reality-forming operation of American exceptionalism. Preemption is an *effective operative logic* rather than a causal operative logic. Since its ground is potential, there is no actual cause for it to organize itself around. It compensates for the absence of an actual cause by producing an actual effect in its place.[32]

In work that maps this condition, it is understood that decision and threat come not solely from large aggregate monopoles of power but also from microscale actors and the imagination of them. There is a continuity here with the discussions of the need to stabilize the "sincerity" of negotiators in the Cold War against the tendency of their equipment to become independently active, to start to generate false positives, to trigger alerts.[33] The decentralization of command and control (established from the 1950s onward in order to avoid the crippling effects of an attack on a central command position) generates an exponential capacity for error and the effects of unintended consequences. In a disturbing sense, the "fear of small numbers" promulgated by the interplay between securitization and terror is a generalization or democratization of the condition of distributing the capacity to launch nuclear weapons to low-ranking and geographically dispersed officers.[34] Irresolvability is in part a democratization of the strategic logics of the Cold War so that it provides the logic for relationships on any scale as part of the infrastructure of feeling.

Cybernetic vocabularies of abstract dynamics paralleling and interweaving with the world of game theory also suggested some antisystemic strategies—the policy, or bluff, of mutually assured destruction, based around a dodging of the constraints of rationality, which also established the lure of going well beyond them. This approach was promulgated by Richard Nixon, who introduced relativism as a basic factor of Western rule through the abolition of the gold standard and engaged in devoted self-monitoring and the tape recording of all phones as a means of correlating the self to magnetic tape.[35] As soon as game theory became a scaffold by which events and evaluations were systematized, it also gained its means of being gamed despite its extensive rigor. Being gamed meant being resolved or broken by the dramatic gesture or having its logic out-reckoned by the imperatives and movements of hidden factors. Technocratic politics is vulnerable to outmaneuver by maniacal exuberance. The durational game of irresolvability itself emerges in the interplay of these factors as the flaw that undermines logic, allowing in turn for its perpetuation.

Such questions drive game theory's convergence with evolutionary models and, in turn, the shift from probability to possibility, the move from the modeling of likelihood to unlikelihood as the zone that must be sought out and mastered. Equally, we move from a state of power over populations based on probabilistic calculation to one based on the composition of variables, where people are acted on via those variables and where the variables themselves become directly operated upon by people as their tokens of interaction that in turn become motions of indifference. It is crucial to recognize that we do not describe one particular systematization of economic, social, or cognitive life that came to fully dominate and thus depleted or actually overpopulated the spaces and conditions of possibility, saturating it with modalities of interaction of a specific kind, loyal to a particular notion of formalism. What is more opportune to trace is the possibility that the multiplication of very diverse conditions and scales in which things become irresolvable has to do with the cognitive conditions of decision-making in infrastructures of feeling and, in turn, the positing of a particular set of idealizations of formalisms, as well as the very material device-like nature in which they become manifest. Part of this is made clear in the fact that the markets, quasi markets, and things that are nowhere near to being markets become things to be interfaced with, taken as the means by which realities must be apprehended and through which they can be articulated as a tendentially dominant set of ready-to-hand heuristics that agglomerate as culture. Cultural theory has attended to these kinds of mechanisms before in other forms; as dialectics, dualisms, hierarchies, structures, and so on. These are explicit formulations of abstractions from the world that feed back into it as modes of ordering and of understanding. What is potentially interesting here is a set of tools that attempts to capture or to abstract from things and processes that are rather less monolithic in structure—nonlinear, complex, inchoate processes that have the whiff of poetry about them as much as that of order. What they have is the capacity for a paradoxical state between emergence and formalism, and expressivity and dampening.

AUTOMATA

Another key contribution of John von Neumann that has relevance here is his work on automata. These are key means for decision-making formalizations to migrate into more widespread and more marginal forms of life, to become a form of life in themselves and to become more multiscalar in form. (Social media and the appification of social interactions are merely one such means.) As institutions, in the economic sense, can, with greater or lesser degrees of interference, be abstracted to sets of instructions, so too can they be produced in programmatic and automatic form. Indeed, given the difficulty of noise in information systems, as Claude Shannon put it, one of the attractive features of von Neumann's work in automata was the idea of "making reliable machines out of unreliable components."[36] This might mean creating an economic and interpersonal logic that what passes for rationality within it would be so readily obvious to all participants due to its transparency that all concerned would voluntarily acquiesce to its operation, independent of their own personal or collective predilections and interests.

It is not that these theoretical objects or logics became fully hypostasized and then imposed on reality to the point that we can no longer see them. Other kinds of restructuring have certainly been made (privatizations; fictional internal markets; criminalization of the activities of trade unions, political activists, and those lumbered with the euphemism of "citizen"; systems of interest, consolidation, and implementation of patterns of ownership; increasing moves to financialization as a fulcrum of "the" economy; a generalization of the equipment and techniques of managerialism, and so on). What occurs is that by these means and others, there is a folding in of certain formalizations into the weft and flesh of life that means they operate as its almost incipient syntax, as an organizational unconscious. There is no "success" of these models, no triumph of the formalism: thousands of them die by the waysides almost as they are offered up as models. Others linger on in half-lives of fading schemata and axioms, but there remains a partial, tendential waterlogging of life with their molecular regime. Such ghosts populate the self-organized political voids that accompany and mirror devastations, generating a wealth of anguish, so that only a few other modalities of "going forward" can elbow in, if we are lucky.

Part of the significance of models such as the market automata of von Neumann is that they are at least nominally capable of sophisticated levels of computation, tending toward a certain kind of completeness in modeling market actions. An important criterion for the success of such models is that they have sufficient requisite variety to capture—that is, to describe, provide scaffolding for, and thus work into—those aspects of life that are beyond the ken of static formalisms. Here, the question of self-organization, and the kinds of second-order cybernetics that implied a high level of reflexivity and flexibility in its experimental models, is worth turning to, as in the following section on the relevant work of Hayek.

HAYEK

Mirowski notes that Friedrich Hayek was not among the number of those political economists who attempted a direct implementation of mathematical formalisms into economic forms but rather that he was someone who promulgated a sense of such forms as necessarily being distributed and nonlinear arrangements. The market, always in an ideal form, is deemed, by Hayek and others, to be the most efficient processor of information regarding price; price, in turn, is argued to be the best measure and means of communicating the value of something in full, across all modalities of value.[37] It is something abstract enough to capture or provide a medium for the articulation and transmission of even unintended consequences. Such use of price is important because it deems it an informational mechanism in which the computing of value is carried out in a distributed and emergent way. At the same time, Hayek also dallied with a liberal form of humanism (although his one stray contribution to the discussion of psychology largely describes systems of neuronal automata), thinking that cognition was somewhat beyond the competence of complete formalization.[38]

In 1961 Hayek attended a conference on self-organizing systems.[39] Later, in a major work, though one not always referred to by historians of neoliberalism, he explicitly proposes a model of the market as a self-organizing system. *Law, Legislation and Liberty* sets out Hayek's figure of "catallaxy."[40] The term is revived by Ludwig van Mises from the Greek, *katallattein*, meaning for both Hayek and Mises to trade or exchange but also to bring into a community.[41] Perhaps "conversion" would be a better term.

Much of Hayek's book is a framework for policy and does not have a particularly philosophical dimension; instead, it presents a set of prescriptions for the role and limits of government, some of which have become grindingly familiar to those who inhabit the world after 1979, the book's year of publication. Since Margaret Thatcher held up a copy of Hayek's *The Constitution of Liberty* and stated "this is what we believe" and since the incorporation of the parties of the center left into that "we," there is little to note but the various toings and froings of the implementation of the program described within it, despite their tendency toward failure in recasting the world in their required grammar.[42] Due to the relatively programmatic and historically specific panorama of the book, there are some things that appear anachronistic; and not solely because they have suffered from the quasi implementation that every ideal suffers from once it mixes with the complexities of reality but because time outstrips them. One thing that does not yet appear so familiar is the radically postnational nature of the "Great Society." Another is Hayek's proposition for a fully self-organized market. The idea of catallaxy appears in the second part of the book. It follows a sustained attack on the concept of social justice as a mirage, an argument that it makes on the basis of principle rather than evidence. Social justice remains, however, a submerged concept that subtends the idea of catallaxy by proposing a form of indirect collective action through the medium of the market—one that would emerge without contamination by notions such as solidarity.

In *Law, Legislation and Liberty*, catallaxy is proposed as the spontaneous peaceful order that is both pluralistic and freed from the end implied by a hierarchy: a state derived from the full realization of the well-operating market that self-organizes to arrange the pricing and availability of all things within the law of property, contract, and tort.[43] Competition is a "discovery procedure" that is also preeminent in aligning the individual with development of his highest capacities by being "kept on tiptoe," being thus both a moral and epistemic mechanism.[44] The economic form of the game of catallaxy becomes the best that can be contingently attained rather than being perfect. Hayek's book *The Road to Serfdom* has state socialism as a coercive system that inevitably generates shortages, with its own characteristic pattern of irresolvability.[45] The games of catallaxy by contrast result in society as a metalevel operation of everyone maximizing their utility through the exuberant search for the greatest margins worked out through the development of a system of markets, or of the market as a system of information circulation that is most effective in establishing a price. Indeed, one apotheosis of catallaxy is its algorithmic manifestation in the contemporary management of financial trading with all the gradients of ascent and descent that this phenomenon affords the humans who are appended to it as an externality and energy source. More widely, as with related writers, the question for Hayek is one of establishing a *procedural* state rather than an end state.

William Connolly makes an interesting rereading of Hayek, in which the latter's discussion of freedom is rendered in more expansive terms and his undoubted feel for moments of spontaneity, creativity, and self-organization are developed, modified, and challenged in ways that, in Connolly's hands, give room to civil society, trade unions, and spiritual and political movements.[46] This is a rereading of Hayek's use of the word "freedom" as if it were not yet part of the contaminated dross of the West's political vocabulary. Creativity is extolled in multiscalar terms and found in a multitude of moments and processes: a cataloging of creativity indeed that explodes the tight definitional channeling of it within solely economic, entrepreneurial, or technical terms. We can say too that there are numerous self-organizing processes that do not correspond to or that actively negate what passes for markets: their strip-mining of creativity, their corralling and thinning out of life, their reductiveness, and their use, as a rhetorical device, to cover oligopolies and the more distributed tendency to systematic expropriation of wealth toward capital.[47]

We must not, however, mistake this model or proposition for the grounding of it as an actually existing market. Nevertheless, each self-organizing system has its own particular consistency, depending on its constitution. In the terms of von Neumann's automata, much of this quality depends on the degree of the complexity of the computation that it is able to accomplish; but such consistency is also established by any of the characteristics of the material forms from which the system is composed and the terms in which it self-organizes. Hayek is interested in the social orders that form as the result of unintended consequences of human actions—the market as a media for trading value, where an externality is the efficient maximization of human self-interest.[48] The market is a metacommunicative frame for second-order communication about the nature of those desires. One of the virtuous results of such a condition is that it induces actors within the market to rationalize their wants: a general degree of rationality is not a feature of economic theory per se but rather a result of the processes through which societies, gifted with rationally thinking individuals, will tend to create the conditions in which it is advantageous to think rationally, thus encouraging the germination and development of the "spirit of free enterprise."[49] Already, we find here the way in which the structure of an idealized market acts as part of a process of subjectivation. In their work, Pierre Dardot and Christian Laval draw on Foucault to show that the neoliberal project is a mode of governance achieved through liberty by the establishment, in various modes, of certain kinds of subjectivation: the production of individual subjects and a grammar of interactions.[50] Such individuals will be properly deterred from actions that do not result in the proper realization of value and so irresolvability is the systemic feedback of what Lars Iyer calls the "normalisation of opportunism and cynicism."[51]

What needs to be added, however, is the recognition that just as human desires are ordered, become rational, and are met through their ordering in a catallaxy, which never fully arrives, so also do other things that are not so neatly described. Effects are indeed triggered and entrained by such a system that must itself evolve rather than simply be achieved. At the same time, there is a concomitant catallaxy of inopportunity, of inertia ordering itself around the first-order catallaxy of the market. What are those things that, as a result, are deemed to be valueless? What are those things that cannot be rationally ordered into a preference list? How do they compose a convex mirror image of the self-organization of nonentities around the all too tangible sizzling of the ideal of the autocatalytic market? These suburbs of ghost towns are where value becomes tangled with the stalling of value, with what is precluded from happening by the incipient logics of that which self-organizes, and from which everything else is a residuum. In catallaxy as an infrastructure of feeling, the sense of irresolvability is simply something to be managed, curtailed, and endured at a local level, by individuals.[52] They are turned back into themselves, locked into their place as generators, agents, and consumers of irresolvability.

What is new here is that irresolvability has been brought to the level of a fundamental axiomatic in social forms and is deeply linked, as modus operandi, to the mechanisms of devastation represented by game theory's integration into the nuclear infrastructure. The condition of impossibility, the strategic trading of annihilation that was and continues to be so blithely and confidently strategized, may now also be seen as a nonlinearly organized state of emergent deterrence of forms of individuation that do not readily align with the ecological ordinates that this condition determines or entrains. We can say that economic markets and systems built to impersonate them may only capture a small set of fragments of the behavioral and data-analytic "signature" of their ostensive subjects. Indeed, just as a drone may misconstrue its target, so too might the mechanism for the entrainment of individuals and the organization of value.[53] The frustration, acquiescence, and rage that accompany this condition are yet to be adequately recognized. Equally, the way in which systematic dumping and pollution have underwritten the economic take-off of a model that only recognizes the automatisms of the market as the form of life proper to growth means that its inverse, in devastations, in anguish, is currently rendered insensible.

QUIDS IN

The outlines of such dynamics may come into view as flickering glimmers just under the sapphire surface of devices or become palpable as a spatiotemporal shuttling between states, as walking into an infinite maze, a landscape composed by the logic of finite automata. Automatic doors, corridors, waiting rooms, themed canteens, tax-free luxury goods outlets, security gates, photo-opportunity points, scanners, ramps, and travelators combine in an endlessly complex and varied manner but go nowhere. In such a condition each automata is but the tumbling, sifting, and resting of nothing, a grain in a slowly moving landscape of sand dunes and rust belts. Particle clouds of diagrams sift and filter each other, like electrons looking for a fix on the edge of an atom. Hot and vast dust clouds of Boolean operators choke on the infosmog of their own exhaustion, engaging in migratory swarming before leaving or entering a city: whispering, *this town is coming like a ghost town*. But such movement may also not blow in from nowhere, being turgidly slow, a depository silting up.

The logical machining of deterrence of the Cold War becomes operative at a molecular level, turns inward, inhabits and implicates processes of subjectivation, which in turn develop their own automatisms to compensate. Being an effective and complex automata perhaps consists of simply extending your skills at playing Candy Crush to as many parts of life as possible. What does history teach us except that life is to be filled with a conveyor belt of brightly colored problems of fitting in and conformance and an endorphin glut fattened out by a succession of microtriumphs from the sensation of snapping, popping, and fixing via effective computing of information? That, and the amorphous, self-organizing anguish of irresolvability.

BLUFF

Bluff and bluffing are a crucial trait of von Neumann's understanding of games. Formalisms were not going to be universal in this model of the world. Instead, his interest followed the extent to which games might create little pockets of explicability, and which might at certain scales have a tendency to be (to use terms proposed by Louis Althusser) totalizing or overdetermining. The bluff—a key element in games such as poker, and more generalizable in different ways via configurations such as theory of mind, the observer problem, bounded rationality, information asymmetry and ignorance—is a means by which such formalisms might be extended. They have consistently been part of the art of statecraft. In being so manifestly present, there is a way in which bluffing is part of the condition of irresolvability, belonging to its systematic intractability and making a political situation so fraught that it is impossible to deal with. Bluffing in the condition of irresolvability becomes a systematic and strategic imperative, one that is tailed by more or less active, more or less ineffectual counterstrategies of conflict resolution. Here, bluff as part of the infrastructure of feeling comes close to the nature of statecraft as fiction.

An attractive proposition would be to make a homology: to state that whereas narrative fiction, especially that of the novel, has concerned irresolvability at the level of the postulation of a psyche or the formal construction of a text, game theory is an attempt to confront and assay irresolvability at the level of political, economic, and military systems via a systematic bluff. This would be too comforting and ready a formulation, however. What we can say is that at different scales irresolvability exists as a condition that is at once subjectival, aesthetic, political, military, economic, cosmic, and so on. At each scale it has its own histories and idiotypic formulations, and particular events occur when these are joined.

There is a complex of aesthetic modes to this condition. Umberto Eco describes one approach to it in the foreword to Nanni Balestrini's novel *Tristano*.[54] Mapping a history of combinatorial literature, Eco suggests that within it, the creative act is that which identifies itself "by intuition, by trial and error, by chance—or by that infinite patience which for Flaubert was a sign of genius—amid the gangue that enclosed and concealed it from our eyes."[55] The particular combination of things that makes a work (incoherent as they are or may be in the way that they hang together), a thought, or a life are already contained or implied as a virtual amid the intermixing of all the systematizations that induce realities. This act of identification, of finding a sequence amid the clattering of combinations, is what constitutes something of art in the regime of probabilistic combination. It is of necessity riven with modes of automatization, and reflects on this condition as part of the work as it deploys it. What does it mean to come into life amid a sea of fields, buttons, templates, algorithms, and subject positions? Balestrini and others writing in a combinatorial mode work a kind of politics on probabilistic governance by, in effect, performing a kind of inhabitation-in-advance of the results of the permutations of chance. If we can inhabit the combinatorial mode with writing, and thus force it to recognize itself as having a poetics, there is perhaps a chance of changing the terms of its composition. Balestrini's figuration of kairos as the choice of *making* an irresolvable narrative is the parallel here but one that answers to irresolvability by running a logic of proliferation, of so many combinations that they cannot be bound in one volume. Each of the printed copies of *Tristano* is a slightly different version. Art too may create blocks of temporality and of the structuring of reason and experience with which to infest the present.

FORMS OF LIFE THAT ARE INCOMPATIBLE WITH LIFE

Irresolvability also has other genealogies alongside that of the strong image of the enemy in a bipolar world as an organizing tension. As Sarah Kane writes, there are situations that "can take away your life but not give you death instead."[56] Life affirms variation but also the means by which a lifelike arrangement of things may act in place of life, provide means of channeling and entraining life, and, provided it is abstract and supple enough, find its power revivified.

For Christa Wolf, in her reversioning of the Medea story, the aesthetic mode of such a condition is what constitutes the necessity for revolution.[57] In most traditional versions of the story, faced with an utterly impossible choice, Medea slays her children and, taking their bodies with her, flees the territory ruled by her husband, Jason.[58] The stakes of irresolvability sometimes imply the impossible choice that can be tabulated and strategized by logic but that have the misfortune themselves to be enacted by subjects, whether or not they are apt workers with kairos. In Wolf's version, Medea subordinates the place of the father by not killing the children, usurping the logic of the father of the faith with its foundation in cruelty. Medea rejects the fateful automata that Kierkegaard's Abraham accepted, seeing it as a masculine trap of the hierarchical ordering of choice and of beings. Summoning magical power, she simply escapes, with her offspring. This, indeed, is the advantage of fiction, a power that we might want to induce in other irresolvable conditions.

Luck

❖ ❖ ❖

Having lived at a site of ongoing radioactive pollution, one child dies of leukemia. Another child from the same site does not. The incidence of cancers being causally related to radiation exposure is understood quantitatively as a risk expressed in percentage terms but it remains unknown which specific individual will be hit. Bad luck. In the lovely morning sunshine, a hefty chunk of ice slides off a roof and hits a passerby on the head. Freak accident. A London cab collides with a fox crossing the road at twilight. Misfortune. A landlady is suddenly inspired to do garden improvements and decides to cut all the branches from a very fruitful twenty-five-year-old cherry tree. Shame. As bees collect nectar from agricultural and wild fields contaminated with insecticides containing neonicotinoids, they become cognitively impaired and their social organization crumbles. Though 75 percent of honey collected across most continents showed levels of neonicotinoids at a level that is neuroactive in bees, its distribution is varied, with South American bees being the luckiest: 43 percent did not get to taste neonicotinoids.[1]

These are the operations of chance: who will be hit, what will draw the short straw to be destroyed or set on the path to annihilation. To some extent these little scenarios echo the thought experiments of utilitarian thought in their allocation of goods and bads. But they are also manifestations of more fundamental scales and dynamics of force. Coming to recognize and analyze these is often marked by a transition from stories to numbers. But there are also more primal figurations whose names we also want to work with here, names such as chance, fate, and luck. Each of these has an ethico-aesthetic dimension that becomes a zone of contestation and carelessness. Ecologies are established in force fields of action whose character, from certain angles, is close to that of playing a game. Species, individuals, genes, and forests are chosen to be "it" via the "eeny meeny miny moe" of power, one that rolls dice and waits on the toss of the cards. It is beyond obvious, however, to note that the die is loaded, the cards are marked. The games' results are audited on the scales of probability, with risks assessed in numbers. The dangers of radiation are known; its risks are calculated and politically accepted as something to be played out biologically. Engineering, planning, and environmental decisions are made. Permissions are granted or are not required; governments are lobbied; court orders obtained; investments evaluated, modeled, and monitored. Chance does not remain aloof from such things.

If part of the role of human progress has been to provide structure against the vicissitudes of chance, it must also be said that the distribution of chance is rarely equitably random. *Ontological load* is the variable exposure to and ability to act upon the conditions of chance in which people, cultures, and ecologies live. Ontological load is the weight of the world as it is forged within political, cultural, and technical structurations of chance. Since control is not absolute, variation in exposure to ontological load becomes a key point of the loading and offloading of chance. This is achieved as much by withdrawal as by intervention, by calculated ignorance as much as by forethought. The following situations, among others, all structure chance: strategically rendered irresolvability; devastations; complex systematizations that acquire insufficient consistency, visibility, speed, or clear causality and that do not always allow for witnessing or knowledge for allocation of responsibility or agency; strategies of risk that allocate bad luck to certain segments of populations or to habitats when, for instance, decisions are made to withdraw support for flood prevention, criminalize or deprecate certain kinds of healthcare, or lace landscapes with poisons.[2] The many ways in which the poor monopolize the greatest proportion of a society's available poverty provides a partial index to this condition. Structuring, fending off, and directing ontological load is the art of war translated into distributions of chance.

Chance is not simply passive and malleable but is also structured differently in different contexts. The score of a football match will never entirely tally with that of a snooker game; the array of chance under Keynesianism is different from that set in play by Potlatch, workers' councils, or certain nomadic barbarians; the texture of indeterminacy is arrayed variably in its many numerical formulations and in the way it is worked with in technologies, architecture, and the formations of organisms and ecologies. Such things are never equal to each other. Chance figures and forces multiscalar compositions that have differing ontological load but also different textures. Figurations of chance do not only acquire aesthetic tonalities as they are articulated in objects as abstruse as theories of probability or as banal as risk evaluations; it is also through the lens of ethico-aesthetic probing and invention that such elements in the formation and propagation of modes of living, of being in crisis, and of advancing toward a range of competing ecological collapses can be understood. There are ethico-aesthetic dimensions to formulations, models, and objects by which encounters with chance are staged. These include figures, myths, and experiences as well as the objectivization, problematization, and management of chance. The operation of chance proceeds through a range of ethico-aesthetic means and figurations of chance run across sites of experience and forms of knowledge.[3] They work and rework it while never releasing themselves from the powers of chance.

This chapter first looks at chance in works by Deleuze and Nietzsche, as well as at their critique by Jean Baudrillard in *Seduction*. Chance here is an ontogenic force. Staged in relation to the background ontological chaosmosis, certain kinds of monstrous accretions of chance occur, monstrous in the teratological sense: driving evolution and the relentless occurrence of events. Within this recognition of chance as a basic ontological force, the generation of styles of the articulation of chance becomes a capacity in itself.

Here, we want to suggest that actualizations of chance produce modes of emergence with distinct ethico-aesthetic tonalities, such as risk, fate, and luck. Risk appears as a form of chance that is prone to being managed, in a manner that is probabilistic, postprobabilistic, mathematical, and out of control. Fate is an archaic, transcendental form of chance, whose forms of explanation lie in waiting for accidents and large-scale disasters that are implicated in the very existence of those undergoing them. Luck is a taming of chance, a domestication, perhaps heralding a self-congratulation on being great, or arriving as a fluke, the grace or viciousness of pure happenstance.

Luck, fate, fortune, providence, destiny, necessity, risk, choice, and other figurations are historic ways of dealing with the contingencies of life, with chance, by describing them. As the conditions of the generation of the new change—preformed by predictions, prestructured (by class, racial, gender, and other divisions), prepackaged into manias and anxieties, preforeseen by politics and capital, or the cunning of a strategy, and reinforced by the offloading of chance—risk, fate, and luck are activated, in new compositions, as fitting modes of inhabitation.

Luck is activated whenever there is a preformation that comes as a narrowing of the future but within which an unlikely beneficial outcome is obtained. It is a form of knowing, explaining, and acting in conditions that could, given different dynamics, be experienced in anguish. When you inhabit a sociopolitical preformation in which the individual horizon of the possibility of economic well-being is defined as much by the chance of winning the lottery as by any other action you might take, to play the lottery is to inhabit luck as a way of participation in the project of society. To get onto the last carriage as the train of job security leaves the economy, to narrowly escape the crushing hand of ethnic discrimination: all invite luck as a mode of dealing with the dispensations of chance. To create a surplus, to generate a new beginning, when what is available is scarce and is thinning out, is to be lucky.

But the dispensation of luck does not always end well. The explanation of luck is used to offload probability onto ecology as an externality or as a site of spillage, to allocate species and systems as the unlucky, whose unfortunate fate is to perish. Fate is thus paired to luck within certain logics. If the Deepwater Horizon spill produces devastation but operates politically as an irresolvable problem, to inhabit it in anguish is composed as embracing one's own fate, and to escape it means to be lucky. Fate is what awaits plants and animals when humans are looking: such is a structuration of chance in ecology.

The activation of risk as a way of dealing with chance is often articulated as probability in contemporary forms of systematization.[4] Predicting and measuring chance with objects and models, tweaking parameters in its composition, allows for another operation on chance. It is a line of argumentation that has affinities with some discussions of irresolvability made in the previous chapter. But while irresolvability is often nurtured and sustained, when turned into risk or chance, it is hoped, it is gambled with and managed away. In other words, chance is to be taught some etiquette.

Luck as well as fate and risk are forms of hypothesis. But they are also a means of explaining or experiencing differing ontological loads. This chapter proposes that there is an ethico-aesthetic to the roll of the dice: a *metis* to dealing the cards that is different with every rivulet of the flow of chance. Amid these, relations to structurations of chance are varied. At the level of the organism, there are structural relations to and incorporations of chance, which are the signature of evolution. At a more immediate temporality, an ethico-aesthetic of fatalism and of the arabesque in plants is discussed in a related way in the next chapter.

In this chapter, after a short discussion of chance, are four further sections, one each on risk and fate, and two on variants of luck. All these cultural figures are means of initiating, understanding, and experiencing contemporary operations of structured chance. There are things that are ordained by structures of economic and political forces, as action on action. But the bad luck of having no chance comes to the fore. To make your home in bad luck takes some doing.

VARIATIONS OF CHANCE

One way into the aesthetics of chance is through the discussion of the "ideal game" in Deleuze's *Logic of Sense*. A classical understanding of games, running through from its formulation in the work of the historian Johann Huizinga to contemporary studies of computer games, entails that one enters the game willingly and that the game comprises "the magic circle," a zone in which the norms of the outer world are suspended, in order to follow through the iterations of logic, skill, and luck inherent to the game.[5] Each game has its own economy of chance and an end point of triumph or loss or cessation of play and refers simply to the constrained range of activity within the circle, the iterations of cards, pieces, or gameplay. One can immediately see the attraction of games and the special dispensation they can arrange from the norms of life by the honing and focusing of particular kinds of sensibility and experience they make possible.

But what is so fascinating in many games is the staging of their partial or full escape from the limits of the magic circle into co-composition with other forms of energy, such as the deep implication of violence within football, and, in a game as serene and mad as chess, the multiple filiations of the Cold War with World Chess Championships (mind games, accusations of conspiracy and manipulations, actual conspiracy and manipulation, vast tranches of propaganda on both sides, and the effect on and conduct of all this by eminent players). The tension of the game is stirred, often to an immense degree, by such things but manages to maintain its gravitation toward the zone of play, governed at times by the addition of surplus rules or procedural agreements covering the staging of the game. The interplay between rule sets and their distributions of potentiality and with other kinds of drives generates scintillating and compulsive tensions that inhabit and stretch the game.

The games Alice experiences in Wonderland are of a different sort, no longer organized around hypotheses of chance but played out in an open indeterminate universe traversed and textured by momentary adherences, prognoses, and gambles.[6] Lost in a delirium of transcoding, rules change, the players become pieces, animals become instruments, the universes of reference and action convulse from moment to moment in passages of cruelty and vivaciousness. The magic circle itself becomes subject to convulsion, fragmenting into an infinitely fissiparous cascade of throws of the dice, at each point of which the dice itself and the form of the throw mutate, staging the flickering between Deleuze's Bergsonian interpretation of the figures of Chronos and Aion or, crudely put, of pulsed time, that of beats, repetitions and refrains, striations, and that of the time of pure becoming, the one shearing off from the other in a dance of pulsions and becomings.[7]

Deleuze's figuration of chance in *Nietzsche and Philosophy* is drawn through Zarathustra, who places chance in relation to eternity through the roll of the dice of the gods upon the tables of the earth and the sky.[8] And in *Logic of Sense*, the two tables of sky and earth have Aion, the indefinite time of the event, as player of the game.[9] Tables act as both the place of the roll of the dice and the place that the dice fall back on, interweaving time as actually lived and as eternal in the generation of becoming. In Deleuze's reflection on the two Alice stories written by Lewis Carroll, it is the dining table and the multiplication table that are placed side by side, but there is no symmetry between the two. These two figures of time, that of construction and that of a plenitude of indeterminacy, interact with the germinations of chance, generating reality, echoing the insight that "ontology is the dice throw, the chaosmos from which the cosmos emerges."[10]

Here, there is a fundamental interplay, following Stéphane Mallarmé, between necessity and chance. The roll of the dice never finally decides things by abolishing chance but invokes the conditions for more entanglements of conjunctions with probability and further limits to infinity.[11] A poem itself can establish a composition of relations between expectation, denouement, and the arbitrary. There is an oceanic fluxion between determination and indetermination—necessity, what arises from chance, is co-constituent with it. In turn, Nietzsche's figuration of chance is always in dialogue with Darwin, a Darwinism not reduced to a system of laws but of indeterminate interactions between ontogenetic forces. Chance moves across, takes part in the composition of, and is transfigured by the idioms of scales, and the polyphony of finitudes and virtuals they impose and imply.

To embrace chance is to put the dice in the mighty cooking pot of Zarathustra and thus to affirm the whole of chance, its rolling and its settling, at once, a lesson or recognition, and not without the condition of reconstitution by further rolls of the dice.[12] As Franco "Bifo" Berardi says in his book on Guattari, "truth must be thought in singular terms, as a gamble"—not simply one gamble, and without inherent rules, but as a condition ramified at each moment.[13] Here, the ethos of chance is one of the open, coupled with the necessity implied by the finality of the expression to be found on the face of the dice when it stops rolling. This ethos is itself born of and reconstitutes the open, in turn disturbed and perpetuating the action of chance, of change upon change. One game, in Deleuze and Guattari, is to multiply the means of recognizing and experiencing the multiplicity of ways by which things occur, and thus to lighten the rigidity of the present regimes of fixes and fixations.

A deformation of another kind, of relation to chance through the play of the game, is the context in which Jean Baudrillard addresses Deleuze's "Tenth Series of the Ideal Game" in *Logic of Sense*. His account begins with an affirmation and intensification of the regime of the magic circle as a place of self-chosen fate that overcomes mere nature. Having a certain resonance with Deleuze's tender elaboration of masochism in *Coldness and Cruelty*, this is an account that is pleasingly perverse.[14] Nevertheless, the grounds for this twist are of a rather different order, as Baudrillard maintains that the multiplicity of dynamics called upon by Deleuze and Guattari are, as a philosophy of desire, rather too readily subsumed within the regime of meaning or ordering, a risk that, to him, is better handled by the cool and measured raptures of a dandy or the explicitly artificial adoption of ritual. Here, relying on a differentiation from the law—configured as nature—there is an emphasis on the game as a choice of arbitrary rules and orders rather than the "naturalness" of chance. The game is solely internal, adopted, chosen, and must be played out—even, or especially, when deadly. The importance of such a measure is that "by choosing the rule one is delivered from the law."[15] There is a reversal here that plays out the argument of *Coldness and Cruelty*, and the aesthetics of domination and release played out in the force calculuses of the asymmetrical figurations of sadism and masochism.[16] Only in the staging of desire can one entertain a release from the vicissitudes of accident.

Deleuze, for Baudrillard, by being so gushingly affirmative of the univocity of chance and being, proposes a species of antimorality, and in doing so valorizes the random, turning it into a good just as, in another manner, science also poses its own technocratic morality, that of the Grand Neutral Aleatorium (a very literal example of which is discussed below).[17] Indeed, Deleuze cites Nietzsche, echoing the Sermon on the Mount: "Let chance come to me, it is as innocent as a little child."[18] Baudrillard suggests that to affirm chance in the way that is done in the *Logic of Sense* is to step aside slightly from it. This is a second-order function that, while still being in itself subject to chance, sets up a reflexive swerve within it, a little turbulence among the lines.

The understanding of luck in the game as operating within the turbulence of chance, the introduction of a social, political understanding, redolent of a certain range of religiosity and erotics, which Baudrillard makes possible, is compelling. It has, however, a relation to chance that is ultimately anthropocentric, even if euphorically tragic as such. That is to say, within its domain of reference, it provides a highly compelling gambit.

The question that a reading of *Seduction* therefore poses is to recognize or inhabit chance, despite the necessary differentiation that any act of recognition requires, is also to encounter the limits of one's capacity of recognition, something celebrated most fully in Nietzsche in his writings on knowledge. Chance, through a million throws of the dice, may produce monsters, fragments of logical or even ostensibly rational order.[19] The suggestion here is that these discussions offer the development of a sensual and political understanding of chance that establishes it as the grounding condition for modes of being and one that is perversely synthetic in its ethico-aesthetics.

One might say that such a line of inquiry replays something of a joke such as the following.

> Pardeep has had a life of bad luck, an atrocious wife, a grinding job, asinine and repulsive children; he prays to God to give him some luck. The chance to win the lottery and resolve all his woes and lamentations. Nothing happens. He prays again this time, really hard. God, please, give me a chance for all my years of misery, help me win the lottery and have a little ease. Nothing happens. Life, or what passes for it, continues in its usual painful manner. Pardeep tries praying for the third time. This time God answers, a little wearily. He says, "Okay, Pardeep, I'll try and sort you out. But give me a little help, will you? At least buy a lottery ticket."

Chance must be prepared, but chance prepares itself.

Given this background layer, what does a sense of the ethico-aesthetic as generative of forms of the *art of living,* or of a *metis* of chance, present? First, there is some useful artfulness in Baudrillard's move toward an embrace of artificiality. Ultimately, given his emphasis on sensibility rather than ontological states, it seems they offer no real contradiction to the wild nature of the true game gestured toward in *Logic of Sense,* offering instead a gaming of such conditions. Baudrillard perhaps asks, what does one do in the context of ontological chance from the perspectival point of one's formation and unfolding? In order to structure the abyss, to carve out agency, to invent new possibilities, chance is to be rethought; plunging into the unknown may tease out some hand in chaos. This is a question worth developing, but additionally to expand, beyond the simple register of the human to encompass the ecological considerations that run through such a scale.

Here, we should attend to the warning in *The Logic of the Worse* by Clément Rosset, who says that chance is impossible to think about, because to do so always poses reasons, some kind of categorical operation that betrays it by fixing it in an armature of understanding that delimits it as fundamental chance.[20] Becoming open to chaos is an encounter with the unknowable, misapprehension itself then adding to the mix. Chance is then doubled by the interplay of nonknowledge, gamings, ruses, and modes of risk; it proliferates in forms of stupidity and cleverness but also in the ecological interplay of forms of luck and fate, as structurations of ontological loads.

It is then the composition of the interplay between chance and its structurations that can reform ontological loads and reformulate appropriations of chance.

CHANCE AS RISK: MACHINES OF CHANCE

Perhaps related back to God's exasperation at the player who refuses to begin the game, and to the question of how one might know if they are a good player rather than a predestined loser, Deleuze in *Nietzsche and Philosophy* suggests that a bad player makes use of several throws of the dice by building the spider's web of reason, mitigating and anticipating fortune.[21] As the structurations of the spider's web of reason extend, its threads also attempt to fold chance inward and make it tractable. Indeed, by virtue of certain experiments, modes of unreason are mobilized in a rationalized manner. Rationalizing unreason is indeed a character of our moment in time. Numerous instruments, devices, and technical infrastructures are elaborated to manage such staging.[22]

An enduring example of such is the Galton Board, a set of pegs or nails set into a board like a bagatelle or pachinko game except with evenly spaced pegs following the dimensions of a Gaussian curve, setting out a distribution of chance with a triangular-shaped profile. Balls are dropped onto the board from the center of its upper edge. Each time they hit a peg, given a further layer of probability determined by variation in material properties, they have an equal chance of falling to either side.[23] In a series of such fallings, strikings, and falling again, the balls have a greater likelihood of falling in the center of the distribution range of the board, with those falling on the outer edges of the range being much rarer. Here we have a conflation of both the notion of chance as a pure force, interrelations with which produce the idiosyncrasies of a particular realization of a specific instance, and of the constructivist notions of chance, in which the fallings are produced by specific configurations of chance as an idea in mathematical terms (as a binomial distribution).

As a device with a variable history of ending up in unsuspected places, the Galton Board plugs reason into unreason through its application in certain long-lived experiments in the use of psychic powers by the U.S. military during the Cold War. Alleged psychics were paid, over several decades, to sit on a sofa in front of a large, glassed Galton Board, watching polystyrene balls bounce to the bottom, the silence and slowness of the spheres contributing time and peace toward their efforts to predict the point at which the balls would end their fall. The point of such attempts at prediction was the entertainment of the possibility that marginally psychic powers might be turned to strategic use. The chance that psychic power could be of use was to be determined by predictions upon an object materially producing a probabilistic distribution: chance captured and transformed in the process of raising it to the third power.

The Galton Board, among other similar artifacts, is a means of both entering into and inhabiting chance but also constructing it, most importantly, through an axiomatic object. It suggests one mode of an ethico-aesthetic relation to chance, enunciated through a mathematical model, a fairly reduced one to be sure but one that, in the rattle and clatter of its operation, vibrates in a way that opens and prepares chance for interaction with other powers chancing upon various modes of action.

Because the progress of the ball's fall is both a specificity of its actual occurrence in the individual descent and something that operates at the level of its mathematical contrivance as an ideational and axiomatic force, it brings into a state of flickering resolution the relation between the event and the problematic it produces, and which the event will later be turned into as the result when the ball lands. History or becoming here produce events that are apprehended, interpreted, and made redundant as results that are also problems. Each fall is unique but is apprehended by the structuration of the problematic, by the mechanism, in and of which it becomes manifest. The different modalities of time intersect here, but we can also say that their interrelation is structured by preformation, not only by the endless rolling back and forward of the dice but by the tables or grid of pegs upon which either falls.

Action today is integrated in a multiscalar way with numerous forms of preformation, not the least of which, in carrying through the relation of reason and unreason, is found in the reversals and enhancements of fortune promised by modeling, risk management, the biopolitical force of statistics, and probabilistic methods to mark out, summarize, and shape chance.[24] Just as the Galton Board provides one route into the understanding and shaping of chance, so there are numerous others, each with their own range of qualities and dynamics, moving across instantiations, and each, as events, opening up new rolls of the dice and instigating the possibility of new problematizations. A plethora of management and assessment techniques, tables, slides and spreadsheets, decision-making aids, and optimization exercises are all artifacts that capture and problematize chance in order for it to be "solved."[25]

While such problematizations may haunt chances, they never resolve them and are unable to drive new unfoldings of the possible without becoming manifest as something more than an unrealized iteration of chance. Here, while the Galton Board has similarities with the ideal state form of hierarchy (with the simple but telling, though perhaps only ostensible, difference of a uniform distribution, where all options of traversal are taken simultaneously), provided, for instance, with "a system of vertical communications—via the region, the district and the kolkhoz committee"—each of the board's transmissions may end in conditions of vagueness or irresolution, or tighten into full stops.[26] Given the perfection of the board, there is not enough happening to make it truly complex. It is left to the matter of dreams to allow the ball to leap sideways and backward or thicken or sleight into new kinds of machining of chance.

Actions on randomness produced by its theorization and dreaming are manifold. In a simple form, from a Russian folktale, a mystic hag appears before King Oleg to prophesy that his favorite horse will cause his death. Oleg sadly sends his stallion away in order to evade the fate set in play by the prophecy. Later, having learned of the horse's death, he travels to the distant corner of the kingdom to visit the horse's dried-up skeleton and triumphantly stands on its skull in order to prove his victory over fate. At that moment he is fatally bitten by a snake that had made its home there.[27] In a more recursive figuration from the seers of economics, Robert K. Merton introduces the concept of the "self-fulfilling prophecy." He describes how, for instance, fears of bankruptcy threaten to produce bankruptcy.[28] The first is a form of fate, the navigation and construction of which we turn to below, and the second a form of structural delirium engendered by contemporary modes of chance management gone recursive by its anticipation. Building on this, in a famous paper on investments, Alfred Cowles and Herbert E. Jones showed that the value of stocks tended to go in sequences rather than in reversals. That is, if they were announced to be going up, they would be more likely to continue going up, and down if down.[29] They warn, however, that forecasting based on this apparent effect "could not be employed by speculators with any assurance of consistent or large profits."[30] The luck of the investor becomes a means of traversing chance via its problematization: its equal entanglement in ordering and preformation.

The actions of the observer in such figurations—such as mimesis, repetition, anticipation, precaution, neglect, and abandonment, whether automated or not—all striate and churn chance. Modern structurations of chance mediate between its problematization and dreaming up, feeding preformation and anticipation into processes of subjectivation as well as into other processes in fields such as financial judgment and organization. It oscillates between the law and the game (in Baudrillard's terms), moves across micro- and macroscales, and becomes ordering patterns that bring together dice and tables; horses', serpents', and bees' fields; roofs, banks, gardens, shale, rocks, and water basins; economic manias and collapses; gold rushes and stagnations, all on a roll. As William Burroughs, in a phrase reminiscent of the probability theory of Thomas Bayes, says, "Now every child knows there is one law of gambling: winning and losing come in streaks. Plunge when you win, fold when you lose."[31]

Here it is useful to recognize the insight of Berardi's work on the modes of alienation that cut through and constitute the modern soul: the shameful to speak of bouts of depression, which are constitutive of the contemporary economy.[32] These are abstract dynamics that move across the myriad synapses of markets and the multitude of choices in a person. Panic and depression are the psychic states, alongside "irrational exuberance," that are particularly accentuated in the present and are among those adequate to the volatility of stock markets and their double in predictability, austerity. A wretchedness of the soul is itself always subject to another bout of arbitrage and hedging. Such woes are tracked, mapped, and tested according to their predictability. Scaled up, as Susan George notes, whole countries reduced to the role of producers of primary goods and held in permanent structural debt to be paid back with the yields of deforestation are looped into a system of obligation and pillage.[33] In such contexts, chance is preformed into a sinkhole, and risk deletes itself.

CHANCE AS FATE

In terms of the ethico-aesthetic structuration and experience of chance, a mode that implies a different form is the ancient one of fate. Outside any necessarily anthropological register, fate is invoked both as a method and an explanation to stage the unfolding of chance within an immediate displacement, a substitution of one state or process by another, an annihilation, an eternal change. A variety of modes of dealing with chance, described above, are employed for those lucky beggars whose fate is to encounter an actual fate. Fate is an archaic mode, fit for those that perish, that are removed from view, that are unknown and become less worthy the more they are affected. This does not impede fate's popularity as a mode of explanation.

An ethico-aesthetic of fate may involve a panoply of instruments. Here, not only do the gods throw dice but humans, insects, and minerals draw lots. Drawing a lot, a conditional object endowed with the capacity to make a categorical judgment—yes or no, black or white, life or death—seems to make, indeed usurp, some of the ontogenetic and phylogenetic agency of those around (a spermatozoid being selected by ovarian fluid or other means, for entry to an egg, destines all the other sperm cells to mortification[34]), while obscuring more complex, multiple lines of actualization. Drawing lots can exist in the form of complex systems: here the selected wriggling lot itself is dispersed into the process of becoming and the equilibrium and disequilibrium of a catastrophe. The Chernobyl explosion can be seen as producing such a chain reaction of order out of chaos out of order out of chaos: an order of catastrophe, an order of the nuclear plant, an order of territory now inhabited by lynx, an order of mutation, an order of thyroid gland cancer, an order of the beta-decay of Pu-241 producing an ever-growing level of Am-241 that will only reach its maximum in the second half of the twenty-first century—where all order is a fluctuation in chaos. Here, a range of scales and arrangements of species and cells, organisms simple and complex, and their future representatives and mutations drew the unfolding complex lot of their fate.

Whereas the Athenian democracy of the fifth century BC relied on sortition, a process in which political positions were filled by a selection process decided by black and white beans being drawn along the candidates' names, the currently official forms of throwing oneself into fate are supposed to be circumscribed by a very few well-delineated spaces and procedures, such as horse races or lotteries.[35] Here the lottery is a very public form of such: the degeneration of an abstract form or decision-making process into banal fate, it is celebrated as an unimaginably glorious prize to the commoner who receives it. A divine throw of the dice adding variety to the mundane to produce another tabloid event, perhaps with a moral ending to come in later coverage. At the same time, more complex alterations and productions of fate are hidden in dispersed networks of incidents, connections, processes, accidents, and decisions. Ecological disasters (such as oil spills) or pesticide irregularities are primary examples here: fateful, they are out of sight; governed by network logic, they have every and no clear point of entry, no white beans to draw. Ironically, in terms of fate, there are only fatal lots.

LUCK AS HOMELY CHANCE

In his meditation on globalization and violence, Arjun Appadurai recognizes uncertainty and incompleteness as a driving force in the generation of ethnic and national certainties, things people hold on to in the context of globalization and something we attend to in the chapter on home.[36] The distribution of certainty and uncertainty across the globe is a crucial means of understanding the composition of the world. Certainty and uncertainty also have different kinds of valence and meaning in different locations for different people, ecologies, and societies at different times. The distribution of certainty is a crucial political question but also a thoroughly experiential one. At the level of the individual specimen, it is often experienced as luck, a run of good fortune: the stupid luck of being born into a particular skin color or into a nation not threatened by the rising ocean levels, the luck of a forest with only small quantities of mineral resources underneath its roots, the luck of a deer able to break ice to get to grass.

Things move from risk to luck and back again, and in so doing they change ontological status. A stroke of luck, whether good or bad, is a domestic form of chance conjoining the scales of home with ecology, a theme of our last chapter. Domestic scale is a combination that measures chance into a form of understanding, that manifests essentially as a belief, a myth, as something subject to magic. Like risk, luck links the divinatory to the computational, as sorcery.[37] As a mythic force, luck depends on belief, even if it is the faulty logic of a belief in the disbelief in the belief in which there is belief: "Certainly, I don't believe that a four-leaved clover brings luck, but I heard that, by chance, it brings luck even to those that don't notice it." Please do not step too hard on the cracks in this argument.

Luck is a means of explaining chance in advance of its occurrence or after it has taken place; in this it mimics the eternal time of Aion. But it is also a form of staging multiple arrangements within which chance can be played; it is a form of energetic and ignorant living through the throw of the dice, a rhythmanalysis of the self, working on the beats of the experienced time of Chronos. With insight, empathy, and effort, the need for luck as a scarce resource can be diminished. The need for the luck of surviving a land-mine explosion is removed by an effective ban on land mines, the application of the precautionary principle. Those for whom reliance on luck is as good as any other measure available, because no other measures are available, are complex figures: accursed, holy, invisible, in some cases, but also repositories, turning points, and improvised devices for bearing and yielding ontological loads.

Such luck may also be subject to kinds of master planning. Luck is found in correlation with technical instruments, precision, political measurements: whether one falls in or out of a massacre of the innocents may rely on possessing documents or may be due to one's dimensions in relation to a rule marked in centimeters. Finding means to meet these arbitrary criteria can be the curse and course of a biography. In economies of debt, being cathected to high or low interest rates are a form in which luck is personalized, a life taken on a joyride by the entelechy of interlocking financial devices.

Luck is not nice. Luck is a factor: in its domestic mode it may take the role of a document, an ordering or systematization that puts one momentarily on top of or ahead of the house's turn, generating the unlucky. Good luck is always fitted with a downward spiral en suite. Play with or against it is never exhausted, unless, of course, it ends.

Luck is an antireason, a superstition that has its own logic, a kind of vaccinatory ruse by which the unreason of chance becomes recursive. Luck is the taming of chance that is replayed in order to enter into a harmony with larger networks, making it safe. As an ethico-aesthetic approach, it is a refashioning of chance to make comfortable sense. It is an unjust form of harmony to be found within the unreason of life, or rather than a harmony, a kind of noncorrespondence between things, an unsympathetic magic. Luck, while being a trivialized form of determinism, is the charming of chance.

It is not quite the case that forms of luck pertaining to nonhuman animals are always bad, or that other animals cannot charm, simply that with humans in a place to observe them it may seem likely that they are so. What bad luck for a fox to cross the road, carelessly leaving itself with a spilled belly as a monument to the unused chance for a driver to release the accelerator pedal. A good harvest makes a lucky year with plentiful food, allowing for the survival of two golden eagle chicks rather than one, thus saving the life of the "spare" sibling, normally sacrificed under harsher conditions.[38] What a good harvest that 27 million chickens are killed every day in the United States, and 66 billion globally every year, and how easy it is to palpate a nervous twitch of outrage as a lazy artist installing the instant scandal of an animal's corpse. Such contexts in turn end as nearly nothing: the breakdown of matter on asphalt or the chance for new viruses, such as H5N1, to breed, given the unspeakably good conditions in the well-ordered mechanism of the battery farm. Leaving nothing to chance prompts chance itself to evolve.

WORSE LUCK

As a forcing pen, the figure of the dice, like that of the Galton Board or the system of lots, is ultimately too granular to encompass chance. Zarathustra, we remember, threw his dice into the roaring cooking pot in order to fuse its fragmentary parts rather than simply affirm their articulation in a well-demarcated branching system. Chance is flowing, coiled, and multivalent, as much as it is also abrupt and fatal. There is something truthful captured in Rosset's observation that to describe chance is to ruin it. But this is, in a sense, to see chance as solely natural. We need a nonanthropocentric sense of chance in order to recognize its ecological dimensions and to understand chance as it intersects with the kinds of chance rendered as formalisms and blindnesses, as structures generated by humans and, in turn, by further structurations, including descriptors of chance. This is something distinct from the games that Baudrillard describes (in which luck is a means of making sense of chance, a way of making chance tailored to the scalar level at which it is experienced), but it speaks of other kinds of systematics, economies, and abstract instruments of chance with their varying forms of concretization and problematics, and how they in turn feed back into the rolling again of the dice.

The ontological condition of chance is necessarily ironic, in that things play by multiple layers of interlocking, fraying law, interpreted in turn as the scalar proliferation of problematics and a humor inspired by such doublings and triplings, the hypnotic stupidity of the depths of possibility. Here we can wager the assertion that it is also possible to invent new modes of inhabitation of chance. Luck, fate, and risk are all forms of such invention, among others, as well as declensions from it. In the present conjunction we are perhaps faced with the challenge of the means of inventing new chances, not simply taming old ones, getting frayed wagers rightly bundled. And it is in this invention that *metis,* the art of composition of openness to chance in relation to its reduction via new means, can be honed. But, worse luck, we have yet to gather a vocabulary of sufficient harshness to attend to the outright deletions of chance that certain compositions associated with our species also seem capable of provoking.

Plant

Plants, and particularly flowers, often stand in as the quintessence of the aesthetic object. They bring together the architecture of charm mixed with an ineffable functionality. Plants embody variation and unfolding, vigor and decay, and provide the gathering point for a catalog of tropes on the passage of time ranging from the ancient, in the case of the gnarliest tree, to the transient, in the budding of the briefest flower. Plants scintillate in the contemplative mode of the aesthetic because they dramatize and embody so many of the contradictions that compose life. While plants are all these things, the scope of this chapter is to follow another line of inquiry into their aesthetic action. Along related lines there is an exuberant growth of writing on and through plants: Michael Marder's *Plant Thinking,* Jeffery Nealon's *Plant Theory,* Matthew Hall's *Plants as Persons,* and, to add fungi, Anna Tsing's *The Mushroom at the End of the World,* among a growing number of contributions.[1] Such work complements animal and multispecies studies and takes part in the widespread attention to what is often figured as the nonhuman in the recent humanities and posthumanities.[2] A broader terrain is also populated by the literature on a cultural history of the significance of gardens and of landscape; by research in biology, which shares problematics with process philosophy; and by work that addresses the networks of relations arrayed around different forms of vegetal life.[3] The proliferation of interdisciplinary domains that might be taken into account in thinking through the aesthetic activity of plants, or in investigating what an expanded sense of the aesthetic that would take the vegetal into account might be, marks in itself the significant way in which plants make themselves felt. They are foodstuffs and co-respirants, indifferent and intimate, activists of putrefaction and regeneration, both our darlings and what will digest us.

Given such a backdrop, this chapter aims to home in on a particular gambit, that of an expanded aesthetics of plants that would take the form of an ethico-aesthetics. It takes as it starting point one of the most interesting tendencies in recent work in botany—that of the question of plant intelligence. Often drawing on prior work in systems theory and cybernetics, the questions and approaches developed in this field are humble, attentive, and inquiring, well aware of the dangers of cosmic overstatement, and in this they tend to achieve much.

Cybernetics describes an entity or a system that makes observations about its self and its environment, deriving goals and establishing actions in order to achieve them. It may reflect upon or adapt its mode of action and indeed change its mode of reflection. Gregory Bateson proposed the term "mind" to describe the information processing capacities of a system such as a biome, something chiming with other cybernetic accounts, such as those of Stafford Beer.[4] This approach is suggestive, since it allows for the detachment of the term from the human or animal brain into a more generalizable consistency. Nevertheless, what it powerfully gains in generality and the capacity to work with interrelations of numerous kinds, it may risk losing in the ability to address the particularity of particular organisms, behaviors, and species.

There are also current calls for the recognition of plant intelligence to describe the complex ways in which plants negotiate their environment, communicate, or respond to each other's state, and change their form or behavior according to environmental conditions.[5] The question of "intelligence" as a broader category is something that haunts the present, with much agitation around its extent, location, kind, and, particularly in relation to that to be developed by computational machines, the extent to which it may suggest and enact relations to the human species.[6] Discussion of it is too often reduced to a SWOT (strength, weaknesses, opportunities, threats) analysis, carried out by philosophers trained to provide bullet-pointed reports and strategic threat assessments oriented around risk and the possibility of a conservation of the unconservable present. Intelligence becomes a list of features that miss what they attempt to address by looking for what constitutes its reduction. Intelligence, thus, can act as a form of absolutism, one that perhaps stands in for and reposes some of the claims made for or on behalf of life, as either terrain for the gross action of simplified entities conjoined in simplified rule sets or for a kind of suprasensible holism.[7] In turn, blurring the difference between intelligence and thinking is perhaps a challenge particular to certain well-trained humans when they attempt to substitute the former for the latter.

The proposition of the specificity of plant intelligence, however, also allows for the detachment of plants from the status of simple passive objects, though it has been criticized for drawing too many connotations of enbrainment down on the heads of sunflowers and the crowns of trees.[8] Some of the more sophisticated arguments for plant intelligence are able to abjure this potential foreclosure when they are more interested in attempting to explore the conditions of possibility of such a problem—explore the definitions of intelligence, which themselves act as a form of contingent proposition working toward something. In short, intelligence in such accounts is excitingly figured as a capacity for problem solving, as embodied, as a shaping of habitat niche, as commensal interaction with other entities in the soil, a means of negotiating the interaction of physiology, resources, and environment.[9] As Trewavas notes, the term "intelligence" itself is equivocal, something that is in the process of being worked on and figured out, with numerous definitions operating in various fields, which in turn proliferate the figure of intelligence as a capacity, force, and mode of interaction and constitution.[10]

Related formations of research address plant communication, sensing, and behavior.[11] Charles Darwin's late experiments with plants—specifically testing the response to light of various parts of a young plant—is often cited as providing a methodological and conceptual platform for such work.[12] Research in this area has, however, only substantially taken off in the last two decades or so.

Communication describes the way in which plants directly or indirectly emit signals or cues that others respond to and that may or may not feed back in turn to the original source of the signal or cue. Plants may, for instance, release volatiles, such as ethylene or jasmonic acid, into the surrounding air, when their leaves are nibbled, which are then sensed by neighboring plants. The roots of certain species as varied as trees, beans, and tomatoes may enter into relations of sensing, nutrient transmission, and signaling via developing complexes of networks overlapping and interoperating with those of particular fungal microrhizoids. Such relations can be highly sophisticated, with their intensity depending on the degree of certain nutrients in the soil, the rate and quality of interaction, and, among other factors, the ability of the network to sustain a multigenerational colony of plants. Here, variably in different kinds of plants, the release of signals or cues makes conspecifics or genetically close individuals alter growth in order not to conflict over resources. An example of this is crown shyness, famously epitomized in the growth patterns of the camphor tree (*Dryobalancops aromatica*) of Borneo, in which the outer leaves and branches of the crown of an individual tree stop growing at a point well short of those of the conspecific trees next to it. Other experiments point to the capacity of certain plants to attenuate their root growth given the near presence of other individuals.[13]

Communication implies some form of mediatic relation between one plant and another. One of the most fascinating aspects of such work is the formulation of the idea of senses in plants. Here, there is a going backward and forward between the figure of the senses in animals to suggest that certain aspects of plants can be described as analogous to these, but there is also a proposition that senses in plants are in excess of those to be found in humans and other mammals. In such an argument, human senses are therefore only a metonym for a much wider range of capacities of interaction with the world and of affect in it. Aside from the ability to send signals to other plants, experiments have shown abilities to sense and respond to the presence of bacteria, the movement or chemical traces of predators, and the placing of insect eggs on the surface of the plant. More generally, it is the ability to sense light, chemicals, touch, temperature, electrical impulses, and sound waves by specific sensors or more diffuse systems of feedback. Much of the range and specificity of plant sensing capabilities is as yet unknown.[14]

Behavior in plants is, on the face of it, relatively straightforward and, in such an example as heliotropism—the movement toward a source of light and the avoidance of shade—is well known. What is still unfolding is the understanding of the varieties of behaviors across species and across the various types of growth embodied by particular kinds of plants. The range of ways plants respond to their environment, to predators, and move toward resources or away from and around blockages is immense in its range and includes the well-known fast movement of species, such as the Venus flytrap (*Dionaea muscipula*), dancing plant (*Codariocalyx motorius*), or sensitive mimosa (*Mimosa pudica*). It also extends to the sophisticated ways in which vines address the surfaces of other plants with tendrils, stems, roots, and inflorescences, and to the way certain plants "advertise" the presence of their herbivorous predators to the species that in turn prey on them.

More recently, there is also a movement among biologists and botanists to make an argument for a form of neurobiology in plants, via the mapping of their electroconductivity in response to internal and external phenomena and the transposing of aspects of such responses in terms of information.[15] Such a claim for a neurobiology is partially interesting because it expands the potential domain of the "prehuman soup" that Deleuze and Guattari locate in the "cerebral-nervous milieu" and sets its center of gravity outside the human and without reference to a cerebellum or a centralized nervous system.[16] (This book also follows such arguments in the chapter on anguish.) There is a useful point of reflection implied here in the constitution of the prehuman and that of the posthuman and the variable movements between them.

Discussion of plant intelligence can also be found in the work of Arthur Schopenhauer, where it forms part of a means to address will in plants.[17] Will here is a force of becoming that can be expressed as a kinetic force (for instance, those of mushrooms that push up a stone slab on the street) or an exploratory force (such as a convolvulus creeper making a slow sweep through the air to find a support for its growth). Nietzsche further draws on Schopenhauer to compare the will to power with the probing motion of a plant as it looks for succor.[18] In related terms, the poet Francis Ponge describes a "will to formation" among plants and also animals.[19] Here, the will is an abstract force of becoming that is concretized specifically in the particular characteristics of an entity. It is a force that resides in an organism that may in turn be composed of subcomponents that have their own wills and tendencies. Each of these becomings in turn has its own characteristics, not reducible to a simple state of change under a uniform condition of indetermination.

This chapter proposes a discussion with these currents but in slightly different terms. What would it mean to address plant intelligence, senses, communication, and behavior as a form of ethico-aesthetic activity? In what follows, we suggest a way to phrase the plant intelligence/sense debate in terms of expanded sensorial processes. Ethico-aesthetics maintains that ecological processes have a sensorial dimension: not reductively sensual but having meaning in relationship to the disposition of the organisms to the world, to the particular dynamics and consistency of composition of organisms and between parts of ecology, and the idiosyncratic rules of such dynamics and consistencies.

This chapter thus answers directly to this book's proposal to think ethico-aesthetics in general relation to ecology. In terms of its conceptual proposition, this chapter has its closest affinities with the chapter on anguish in developing another inflection of a way of thinking aesthetics in relation to an extended sense of ecology. In terms of its mode of working, the chapter is closest to that on irresolvability. Both address the questions of how compositional dynamics retain consistency across scales (uniting, in case of irresolvability, for instance, those of self with political doctrines of nuclear deterrence), while embodying and making change (a question to which cybernetics and systems theory previously tried to provide some answers). What is the consistency of a multiscalar thing or process? Both chapters ask what the forces are that come to produce and bind together the scales of composition pertinent to plants or to irresolvability, which could be as varied as chemical reactions and the formation of social conditions, logical formalisms, and the industrial-scale production of fear, each with potentially many scales operative. Manifesting and having significance differently for each scale, irresolvability or the ethico-aesthetic powers of plants forge varying compositional consistency at scales of formulation while retaining the capacity to move scales.

Equally, the composition of abstract dynamics includes the question of indifference, intractability. For example, the sound waves of human screaming are irrelevant to the composition of freezing water. Yelling at freezing water will not influence how water molecules interact with each other at the point of freezing. One might experiment with using water as a form of recording media and devise a reading and playback device that could extract vocable sound from such a media, but the chances of empirical success are slim, to put it mildly.

Finally, this chapter addresses the question of intelligence in plants via an ethico-aesthetic mode that appears at various points across the whole book. The "cunning" of plants is for instance affiliated to the art of inhabiting chance in the discussion of luck. To embody and thus inflect the flow of chance in a way that produces a favorable or interesting outcome is *metis,* an art of living with other forces that attempt to structure chance. Kairos—the ability to work with conditions in which one is, to steer and be part of the river of fortune that one is immersed in—is not only open to human experimentation but is also actively composed by plants as well as other organisms and ecological processes as they are both constrained and embodied by it. Here, we are interested in plants' ways of inhabiting an aleatory planet. It is a form of kairos that runs across their compositional consistency at multiple scales, and in various idiomatic forms as part of wider ecologies.

ETHICO-AESTHETICS

Mikhail Bakhtin's formulation of the ethical and the aesthetic works in part through a recognition of the fundamental plurality of interactions that constitute aesthetic and cultural forms and processes. His work is a useful preparation for addressing plant intelligence, sensing, and communication as a form of ethico-aesthetic due to its predilection for translation into a means of tracing these things through multiple ecological interactions and their specific manifestation in particular organisms. In Bakhtin, the deep historical development of language; the play of jokes and phrases in everyday cultures; the movements of folk, popular, and mass cultures as well as their articulation by a particular writer and a certain text; and subsequent cultures' rereading and vivification of them all constitute active forces in literary development.[20] The scale of such a conception of literature is one that is massive and open-ended, but it is also one that is able to articulate miniscule interactions of entities and processes with precision and attentiveness to difference. This is in part due to the fundamental role that the notion of the dialogic plays in his work: Bakhtin emphasizes a play between the insides and outsides of a thing. Such insides and outsides might include those between scales such as the levels of a text, modes of reference, a noncoincidence of address and response, and so on. In doing so, it constructs a cultural current as a set of multiple points in which a work of literature or a cultural process takes place. The author's relation to the figure of the hero is but one of these.[21] The interplay between the understanding of the self and of another, of the forces manifest and latent in both, is one that is especially important for the kinds of literature that Bakhtin was interested in, but it can be seen as a particular instance of a wider way in which he offers us a sense of history emerging through dialogic plurality. The deep ontological coursings of reality are always inflected by the individual moment, the person, the phrase, or the movement of an idea as it comes into words. In this emergence, the richness of the ways in which approaches that are differently scientific, philosophic, literary, or erotic and which also lie in popular cultures are able to address, incarnate, or express the composition of reality through massively mutual interactions of dialogic pluralities is a fundamental grounds for experience and enquiry. Concomitant to this condition in Bakhtin is that nonknowing; alienation; and the multiscalar concatenations of limits, capacities, and forces, amid the sensations of wonder and allure and the gradual patching together of contingent knowledge, are all mutually constitutive of this plurality. The ultimate unknowability of the other for him is thus a motor operating via differential pressures and turbulences; yet this unknowability is not, in turn, something that places absolute limits on things operating at other scales but rather a refusal of the possibility of subsumption. Texturing the unknown are numerous concurrences between words, authors, heroes, and worlds, which perplex or make uncanny a simple eternal withdrawal of a thing into any purported solipsism.

Equally, there is in Bakhtin a sense that each act is an event in itself, with its own traits, histories, and singularities in the trajectory of a "once-occurrent being that is unique and never repeatable."[22] The ethico-aesthetic being is not simply concerned with a mode or system of perception and appreciation of sensory data and its recomposition into a sensorium but is also a consistency of existence and a mode of putting that consistency, habit, or regimen into question through the manifold qualities of its internal and external outsides. Bakhtin thus offers us a language with which to talk about complex phenomena and their movement through history. Guattari draws upon this material to discuss an aesthetic entity as something that first operates by means of a relation of striving to achieve autonomy from its point of genesis in an author and then moves into the multidimensionality of being that is at once social, ecological, mediatic, and aesthetic. It has its own physiological, emotional, and sonorous factors; it keys into registers of previous, imagined, and concurrent acts and in so doing becomes something like a partial object, inducing relations with its surroundings according to its constraints and affordances.[23] Within this approach the ideas of Bakhtin were used to talk about partial objects and the volitional relation between insides and outsides, between wholes that are only completed in relation to another. An ethico-aesthetic reading of plants brings the ecological dimension of such a reading further to the fore: it finds its manifestation through multiple kinds of organism and process, being rendered truly polyphonic.

In this relational movement that also extends to the transitions between the ontogenetic and the movement of learned or on-the-fly actions, every plant writes its own ethogram. By these we mean the unofficial ones as much as the guidebooks to the state of the entity drawn up by professional guides. They are inscribed on the walls of buildings as a branch or a stem's dialogues with the breeze bringing about scratches on the surface, the curves in sand dredged by a leaf of grass, or the time and motion studies scraped on the roofs of buses as they scuff against a low branch of a tree but also in the movements between plants in relation to space, light, nutrients, and chemical interactions between species. The nature of this ethogram is idiosyncratic to each plant and its context.

Here, there is a coupling of capacities of action and power with modes and affordances of sensation. It is not by accident that the model for the arabesque in art is the undulating line of the creeper. The arabesque (more fully attended to in the second part of the chapter), a leitmotif of design from Art Nouveau to parametric design and so on, is an insistence on extension, a pouring outward and a relation to the open. But we can also say that it is a site of tension. It is the demarcation point between internal and external forces, which, at the scale of the symbolic, may be deployed as demarcator of the implied and intended mastery of the organic.[24] This tensile condition is part of what stabilizes and induces the arabesque as a line through culture and which also gives it its power. The arabesque can be taken as a mode of abstraction that is able to move across iterations and instantiations as an abstract dynamic. The ethico-aesthetic is in part to be found in the means of attending to and enacting such movements.

Grasping a sense of the ethico-aesthetic in expanded terms allows for the recognition and creation of compositions that also expand into the political, subjectival, and technical. An advantage of such an ethico-aesthetic approach is that it allows for reflection on scientific accounts of plants alongside other modes of description and involvement in plants, such as the poetic, philosophical, and painterly. All of these are modes of composition as well as translation, which occur as actual instances in relation to vegetal life.

In botany, plant systems can be understood at numerous scales, including those of molecules and systems, plant tissues and organs, whole plants, intra- and interspecific relations, and ecosystems. Each of these scales is arrayed in composition with those "above" and "below" as well as alongside it, producing dynamic arrangements of interpretation and formation, with crucial aspects of the organism, such as the itinerary of energetic and nutritional transmogrification in photosynthesis, moving across and organizing the interrelation of such scales. There is a dialogism here, to use Bakhtin's terms, but also something that is drawn out well by Isabelle Stengers in her meditations on acting in the middle of things rather than adopting the allure of either reductions or holisms.[25] Keeping holism at arm's length is simple perspectival modesty. (The amongness of anguish described previously is a related description, one that in some cases transfigures into the outwardness of home discussed later on.)

To recast this question of acting in the middle: any system or complex of aesthetics, or any aesthetic mode, is revealed in part by its means of probing into, opening up, or seeding the constitution of modes of composition at the same time as it founds them. A state of partial or temporal outsidedness, one of differentiation that plays across the inside and outside, drawing material across scales and enacting transformations, is a precondition for any ethico-aesthetic act because this revealing comes partly as revelation. Acting in the middle is thus both temporal and procedural and changes the condition of knowing implied at its outset.

Acting in the middle is complex, and occurs in multiple kinds and via multiple vectors of interest. Alongside the way in which plants manifest ethico-aesthetic forces within and among themselves, they are also drawn into different kinds of composition. In this chapter we are interested in aspects of plant tissues and organs, and in whole plants. This scale of interest partially arises out of perspectival conditions and is limited to what is observable to humans. The generosity of creativity at the level of the genome would be another route into this topic but has a different, and also fascinating, set of compositional forces at its core. The patterns of light that set up a pullulation of chloroplasts, painting starch in the leaf, are conjoined with ways in which plants become factors in other forms of calculus—middling should also stand in for mediation. In one form of this, in analyses of urban systems, plants may be said to provide "ecological services" as a form of vegetal labor. In such approaches, trees are calculated as a basket of values based on factors such as the storage of carbon, the amelioration of air pollution, the absorption of rainwater that they hold back from drains and sewers in cities, and the provision of relief from temperature extremes.[26] Plants may be treated as a form of engineer, holding terrain together via avid root systems, as in the case of marram grass (*Ammophila*) on sand dunes. They may be placed, according to strict budgetary allocation by an outsourced gardening company, on the grounds of a campus, where beds of low-maintenance plants are to be found solely at the intersection of the most-viewed spots in the transit routes between car park and office. Inside the office, leaves are to be sprayed with water and polish to leave a natural-looking long leaf shine and are to be mopped with a cloth according to an annual schedule. In terms of their literary and rhetorical services, where they are plumbed into systems of words or feelings, plants may be said to be mute, voracious, naïve, hesitant, innocent, invasive, greedy, or ideal. They may be used as portents, symbols, organizing principles, abstract dynamics in fleshy form, to provide the grounds for a naturalism or a recourse to an ideal.

In what follows, we want to elaborate grounds for two possible readings for an ethico-aesthetic of and in plants, and to offer two terms for ethico-aesthetic vocabulary: fatalism and glory.

FATALISM

There's something odd about plants, something indifferent—a quality that should make you take care when you water them. Plants are cold like that, making them, as in many horror stories, apt to inspire an exhilarated creepiness. Plants may also be, as Nietzsche noticed, driven by the moist heat of the tropics, which removes most of the obstacles to variation, and exist in an exuberant annihilating indifference. Plants echo the relentless sun, transmogrify its rays, but they also map the decay it represents. The mediation of entropy that occurs in the gradual, years-long collapse of a willow into the slow unkempt stream that feeds it and undercuts it, and for which the tree provides the sprawling bank, is a toppling that takes an age in itself. Each plant in turn negotiates, in a form of vegetal pragmatics, the relation to its ecological condition, never utterly fusing ends and means in being—instead, playing them out idiomatically, according to species, habitat, the ecological interaction of forces and local conditions, operating at a multiplicity of scales.

Fatalism is a bad thing, a form of resignation, of negligence expressed in the task of composing a self, something gloried in by fanatics, introducing illicit supinity into the personal fitness regime that is supposed to manufacture self-affirmation. One sinks into a stupor of fatalism not only by accepting the world as it is but by inducing it into a worse state by the simple means of torpor or, alternatively, by perpetuating its indefatigability by brazen testing of one's fate. Fatalism is embraced by the relentless rhizome, the ginger of plagiarism, the iris of boredom—such a plant does not simply proliferate but goes on and on.

In plants, fatalism is a mode of embracing the interplay of determining forces: as a leaf tirelessly plays with the wind. Vegetal fatalism is not an automatism producing a self-same—there is always differentiation: among trees, blades of grass, potatoes; in the pattern of growth particular to bladderwrack or the fine genesis of abscission of leaves and branches; in the hungers for nitrogen, carbon dioxide, and other nutrients that always turn an organism to its outside.

Such a compulsion can be found in the drifting Chihuahuan desert plant rose of Jericho (*Selaginella lepidophylla*)—a tumbleweed that rolls in the wind, scuttling along the ground in the breeze in a gray dry state, producing spores and becoming a spiral of green in the occasional moments of saturation with water. What is magnificent in the plant is the way it hurls itself forward, into the air, down into the ground, across and into water, in the turbidity of the ocean, and into the cracks of parapets and paving.

There is a relentless severity in its growth. Plants hurl themselves into and as the world, but fatalism also requires glimmers of reflexivity, something already in excess of fate, some kind of figuring out—and here, in this interplay of scales of composition, it is already beyond itself. If we are to draw out a consonant line of composition in a human life story, the protagonist of Jaroslav Hašek's *The Good Soldier Schweik* is good-humoredly fatalistic in attempting to place himself at the service of the glorious Austro-Hungarian empire. It is not his fault that things go awry over and over again; such things simply precipitate around him.[27] For Nietzsche, plants already know "cunning," an ability to take a certain distance from "truthfulness" and to engage in error, that of individuation.[28] Fatalism always makes matters worse, that is, it finds a means within a condition to, within certain constraints, release an unexpected potential. Here, we can say that it has something in common with the "bad rhythm" of the poet Alexander Vvedensky, where the right choice in composing a poem is always that which is off, slightly out of whack.[29] One of the themes in Vvedensky's work is the nature of things unfolding in time and the chance to catch time itself off-balance. Here, there is a chiming with Bakhtin's formulation of the difference within an utterance between its plan or the intent behind it, and the realization of the utterance—things that may vary significantly and tellingly.[30]

In the work of Daniil Kharms, fatalism is found by addressing oneself to all the things that should not happen in a narrative. The plot switches from a predictable trajectory, affected by the fork in the road of a minor detail. A character who seemed to be the protagonist disappears. A heroic act is undertaken and then found to be worse than meaningless, inconsequential. There is capacity in language to name something extant, and then, by working the categories implied by such a naming, take its existence apart piece by piece. A spider plant in a forgotten corner of a room produces a plantlet, which achieves nothing in particular, so it makes another. In so doing, repetition reaches a kind of stupor—the plunge into which is required for cosmic enlightenment. Kharms has Pushkin and Gogol collide with each other over and over.

> ... PUSHKIN (*getting up*):
> Hooliganism! Complete hooliganism!

goes, trips over Gogol and falls down
Damn! Again over Gogol!

GOGOL (*getting up*)
Complete insult!
goes, trips over Pushkin and falls down
Again over Pushkin!

PUSHKIN (*getting up*):
Damn! Veritable damn!
goes, trips over Gogol and falls down
Over Gogol!

GOGOL (*getting up*)
Shit-trickliness!
goes, trips over Pushkin and falls down
Over Pushkin![31]

The sheer fact of this stupor, the delight of variation within it, always threatens to draw the conflict to a close. It never does. Surely, Pushkin and Gogol are still falling over each other, collapsing onto or dropping ladders on each other in some virtual extension of the text even now. Slapstick is a fatalistic force of becoming. Other depictions of the trajectory of life that occur at a more sedate or illusorily serene pace are a thin parody.

At the same time, we can also see in these narratives a basic skeleton of all stories operating within the medium of text, where, despite the struggle against such a constraint that constitutes part of the history of literature, one thing necessarily follows another. Just as plant species, such as rice or wild tobacco, have a great number of chromosomes (more than does the human) endowing a great capacity of redundancy against the exigencies of habitat, there is a sophistication here wherein the expression of diversity is implied by resolute simplicity, one that is threaded as a movement through the phase space of possible tales. In order to trace such a condition, one has to maintain a certain patience. Gregor Mendel's, Muriel Wheldale's (and others associated with William Bateson), or Barbara McClintock's style of research had a model of rigor that mapped combinations, repetitions, and variations in a way that gave glimmers of underlying genetic mechanisms.[32] Such work requires a taste for fastidious adherence to the regimes of repetition (of well-recorded pollination and seeding) to catch the dance of mutations operating at several scales removed.

The "margin of indecision" that Italo Calvino finds in the process of translation of a sentence or phrase from one language into another is the activity of an "interfering swindler," an interpreter of a reality between an inside and an outside, creating filtered conduits between them.[33] To transpose this condition to that of plants, we can say that the translation of movement into nutrition, soil into starch, moisture into a signal, without so much as approximating a language or a code, is done, for instance, by the working of a root, in a space saturated with potential movements not taken but anticipating those that will be. Here, text works its way through a space of letters, phonemes, words, sentences, ideas, and stories; through the muddle of references, characters, figures, and images; into narratives, systems of composition, economies of publishing, forms of books, oeuvres, movements, lost texts, and imagined libraries. There, roots probe their way through soil, drawn by patches of damp, the edge of a seepage, floods or thin traces of nitrates, working their way around or flummoxed by stones, pressed up against thick drifts of clay astray in the idyll of the loam. Here, pebbles, sand, and schist of sentences get in the way and induce thin tendrils to explore their either side.

Catachresis, a semantic error or misuse, induces variation, a new sprout, a luck-infested roll of the dice. We are not saying that plants are like language but that language may be vegetal—or better, that there are abstract dynamics that cross between them that we have yet to trace. Nietzsche (who often refers to man as a plant) reads error, the space probed by such exploration, as Dionysian, as the variation that Darwin ascribes to evolution. Indeed, "the force of life wills error" and provides the grounds for thought.[34] What fatalism does, however, is to make such a move by indifference, to imagine whatever comes into being—there is little in it either way, except for the conditions of life itself.

Fatalism is also a relationship to growth that may be expressed in a slowness beyond the life span of human civilizations. The Namib Desert plants that may live two thousand years exhibit such a condition: the small *lithops*, a fleshy stone that tugs itself into the ground, or the larger

Welwitschia mirabilis, whose two leaves grow and split in a ragged wig for the submerged stem. Elsewhere, the plant known as king's holly (*Lomatia Tasmanica*) is a clonal reproducer that appears to have been in a state of persistence in a small patch of Tasmania for at least forty-three thousand years. Fatalism here is, like the scientific ethos of McClintock, Mendel, and Wheldale, a state of endurance. Francis Hallé reads the spans and pulses achievable among different kinds of plant temporality as rendering animal life a quick blur, a burning up of time and tissues that fleet past the trunks of trees in their extensive being.[35] Fatalism thus implies a relation to time. One of Hallé's other contributions to botany is the study of the architecture of vascular plants in their relationship to gravity, a subject to which we turn next.

GRAVITY AND A TASTE FOR ZINC

Fatalism is in part the substantial accommodation of something that must be yielded to. It may be found, for instance, in the creative and generative response to what are rhetorically configured as laws, such as that of gravity. Indeed, one of the ways in which we might speak about an ethicoaesthetic of plants is via a reading of recent work in the capacity for sensing and responding to the direction of gravity in plant roots.[36] Such work has its precedents, of course. The pioneer mathematical biologist D'Arcy Wentworth Thompson's 1917 book *On Growth and Form* was shaped by his great interest in gravity as a force in the development of organisms, where, for instance, he famously analyzed the structure of animals by means of the mechanical stresses they bear.[37] Morphology here results from the interaction of the inner resources and organization of the organism, articulated in turn at numerous scales, from the molecular upward and fueled by what it might abstract from the environment, and its dynamic and unpredictable negotiation of gravity. The field of biophysics arises from the study of such forces and formations with that of living evolving matter.

Work on *gravisensing* in plants is partly enabled by the capacity to study plants in the near zero-gravity conditions of space.[38] It also builds on the wider contemporary move away from the Aristotelian and Linnaean tendency to view plants as passive and mechanistic and traces a lineage to

Darwin's later experimental work on plant light sensing.[39] Some of this study is involved in a wider set of arguments for a renewed understanding of the complexity of plants, involving means of collaboration, defense, and cunning.

Plants have been adduced to sense gravity in two ways.[40] One is by means of statocytes in the head of root filaments—small capsules of gas surrounded by a membrane, whose internal concentration is modified by the gravity gradient as the plant moves. The other is by means of statoliths—small starch crystals that move inside the root tip. As the root tip moves through the soil, it encounters barriers or changes in the quantity of nutrients to be found. The statoliths allow the plant to sense which way is downward, since they are heavier than the surrounding root tissue. Video footage of microscopic statoliths moving around within the root tip shows the miniature crystals within a foraging root, navigating between cells, drawn downward by the tug of gravity.[41] Statoliths provide a kind of ballast for this roving root: tumbling slowly between cells, they nudge the root downward, encouraging the allocation of more starch to the points of outer curvature and encouraging growth via stretch-activated ion channels in the plasma membrane of the circulatory system, and triggering the transmission of information about the growth via the release of the hormone auxin.[42] All this happens within the regime of geometrical necessity put in place by the kinematics of cellular and organismic growth characteristic of a particular species.[43]

The statolith crystals provide an inner quasi geology to plants, something that binds them to the earth, but also marks their capacity for differentiation from and within it. Ponge writes about the rock being the boundary to and basis for a carpet of moss.[44] The unrelenting nature of the stone gives rise to a scuff of rhizoids crisscrossing its stupefying surface, making a complanate mat that builds itself upward. The relation between plant and crystal or stone is complex. Some plants build up high levels of the mineral mica in their tissue in order to inhibit chewing, wearing down the teeth of most animals. Others use it to cantilever off at wild angles, like a buddleia sprouting out of a cracked parapet.[45]

The production of crystals by plants is something that seems beyond the capacities of their tender flesh, but in fact the negotiation of complex relations between their tissues and the presence, and incorporation, of minerals is a common problematic for plants. Those growing in extremely salty environments, such as the marsh samphire (*Salicornia europaea*), are particularly prone to such a condition.[46] Its asymmetric storage of solutes, its relatively short root length, and the presence of mechanisms for pinocytosis allow for small-scale negotiation of particles, and the storage of inorganic material in vacuoles. There is a folktale that marsh samphire grows a salt crystal at its head in order to concentrate and gather fresh water, which then flows down the stem to the root when the crystal dries out in the sun, but the real operation of the plant seems instead to be one of internalizing such a mechanism at multiple sites in its body and at much smaller scales. The fatalism of incorporating one's enemy by placing it at the crown of one's growth or the deadly labor of working roots into cracks is here radicalized by turning it into the mechanism of self-incorporation.

The ethico-aesthetic relation of the vegetal to the mineral is extended to metal in Mel Chin's now canonical work *Revival Field* (1991–), sited at the Pig's Eye Landfill Site, St. Paul, Minnesota. A collaboration with Rufus Chaney, an agronomist with expertise in the uptake of toxins by plants, the work sets out a demarcated terrain in which "superaccumulators" are planted in a geometrical form. The particular stretch of land is tainted with zinc, cadmium, lead, copper, and nickel, among other materials. These metal particles are produced by their smelting and end up, via the air, in the soil.[47] The project operated as a testing site for the uptake of particular metals by a range of plants, with an emphasis on those able to take the metals into their leaves and stems rather than their root, making harvesting easier. Specimens of plants drawn from mine sites in Europe were imported in order to start from a position of highly developed tolerance. Once grown, the plants are turned to hay and then burned, producing ore in a manner characteristic of historical uses of plants known to concentrate certain minerals.

Among the plants used in the project, pennycress (*Thlaspi*), a kind of brassica growing in clusters, turns out to be particularly good at remediating cadmium and zinc. Prospectors have historically used it to identify alpine sites rich in these metals, since it flourishes in their presence. The time scale of its incorporation of metals is slow, taking several years, but selective breeding and engineering based around the identification of appropriate gene traits is being probed as a means of intensifying the speed of this process.

Chin relates the operation of plants to more traditional operations of art on metal and stone, such as casting, carving, and reduction—the removal of materials in order to make a shape.[48] Here, there is an atomistic relation to a dispersed cloud of metal particles, accumulated in the soil over years of industrial production. The plants operate as the agents of reduction, reversing sedimentation and seepage. Having historically worked with the concentrated forms of such materials, aesthetic operations may now be carried out with explicit recognition of their resultant diffusion. The removal of materials is no longer only a result of such things as the operation of hand and eye, mallet and chisel, but also takes place through the slow action of root and soil, light, plant, and irrigation. The aesthetic becomes one of decentralized transmogrification, where cleansing, digesting, and concentrating are also dispersed among a multitude of organisms, something that in turn remains unseen until further transformations occur. Crucially, the work synthesizes such activity along with significant technical and intellectual operations in order to articulate this powerful capacity of plants as agents of the recovery of soil to conditions where it approximately correlates to the health of humans. One may imagine other kinds of balance. Such reflections could work in the direction of an approach similar to that of the pharmakon in culture, of the medicinal absorption of what poisons you as a means of warding it off and working with its power, but it can also be seen as a testimony toward sorting, differentiation, processing, and the opening up of internal pockets of strangeness within an organism.[49]

ARABESQUE

A movement between fatalism and glory can be traversed in the articulation of the arabesque. Each of the following familiar plants—ivy, wisteria, or honeysuckle—have their particular patterns of coiling out into the air in search of a support or a source of nutrition. Each of them describes a particular arabesque, a coiled repetition and scrolling pattern that entails the extension of the body of the plant to its furthest reach. The arabesque may be fundamentally modular, as in Islamic tiling patterns, or developed in choreographer William Forsythe's notion of the hyperextension of gesture in ballet, where a movement from the classical style is pushed to an exploration of its limit, but it always implies an interlacing of inner and outer forces and capacities, woven in among others.

The soft convolvulus grows faster than a blackberry bramble and may often be found on patches of "waste" land, tangled around the latter's thicker curving stems as they, in turn, thrust out into the air. Here, there are two arabesques, typical to each species, but one incidentally involved in something of a slow game of annihilation of the other. The touch of the inner surface of the convolvulus stem—as with peas and beans, among other species—generates the curve of the tendril. The cells on the side of a tendril, stimulated by touch, slow down their growth in order to wrap around a support as the nontouching side expands more rapidly.

The characteristic form of vegetal growth in Jugendstil or Art Nouveau—exemplified in the swirling undulate dance of Loie Fuller, who took centrifugal force and the fine modulation of circling waves acting on swirling fabrics as primary force of composition—is an aesthetics of becoming and of uncertainty, something epitomized in the cunning of the plant that perhaps most embraces it.[50] The dodder plant of the genus *Cuscuta* is a splendidly avid parasitic vine that ditches its root structure once nutrients can be sourced from the stems of other plants and is often cited when it comes to recognizing the capacity of plants to negotiate and differentiate between resources in a limited space. It responds to the scent of favorable prey plants with rapidity and ferocity but in turn can live for only seventy-two hours without inserting probes into the stem of another plant to extract nutrients.[51]

Another way of tracing the arabesque as articulation of the aesthetic through the conjoinment of multiple scales is via Francisco Varela, Evan Thompson, and Eleanor Rosch's formulation of enactivism.[52] Here, perception guides and is woven into action; recurrent sensorimotor patterns enable action to be perceptually guided according to the conditions of possibility of perception in the species or organism concerned. (In humans and others, these give rise to functions associated with cognition.) The plant's structure and its sensual and communicative actions live in the immediate present, but as with any organism, this present is coiled into that of many others and into a negotiation of historical change through the conditions of age, nutrient resources, structure, habitat, and others. The arabesque is the most visible tracing in action of that process. What may be read as a mere curvaceous line, the arabesque, is rather a multidimensional incarnation of forces and capacities.

GLORY

In 1950 and 1951 Richard Hamilton, the exemplary cool surveyor of late twentieth-century life, developed a short series of small pictures made with etching and aquatint, including *Microcosmos: Plant Cycle*, *Heteromorphism*, and *Self-Portrait*, which were included in the exhibition *Growth and Form* in 1951. At the outset of Hamilton's work, which throughout explores relations to technology and modern modes of perception, is a trio of works that addresses plants and, with a nod to Wentworth Thompson, the means of technical description of life forces, movement and becoming, including the fecundity of even the tiniest germinating thing. The composition of *Heteromorphism* is simple: a few scraped lines are clustered together, which manifest a seedling, root form, grass or seed pod, a spiral, a fuzzy spiky ball. It might be a botanical specimen excised from meadows also roamed by Paul Klee, except that it is sparser: at the fore of the image, in black ink on cream paper, the scraping of the tools on metal are held in tension with the rigor of the image. In *Microcosmos*, a horizon line gives us a sense of depth as six highly schematized and skinny plant figures grow, lean at odd angles, and jettison what might be flowers or seed heads against the glowering of a patch of black ink skimmed in the background. *Self-Portrait* is a kind of image one might expect to see from Giuseppe Arcimboldo but after the advent of the microscope, with abstracted plant and animal forms arranged on the page to incidentally compose a face from volumes, lines, and voids—one where vegetal form mixes with the geometrical and with the sketched.

In all of this series, there is a sense of fragility, alienness, and movement. These are presently sparse microscapes, but there are logics of growth and composition that the space of the paper can anticipate becoming overrun by. What may be a seed, a polyp, or a migrant from a painting by Jean Arp glides delicately by. The forms grow and exist, but at present they are hesitant: this is probably only the start of something—what Hamilton points us toward is a tentative, but very present, form of glory.

Johann Wolfgang von Goethe's account of the morphology of plants presents us with a metaphysics in which the leaf is the primary organ or *urphenomena* of the underlying shape-making properties of the plant.[53] Each component of the plant is but an articulation, in relation to the environment, of this underlying protean leaf. Contemporary science has perhaps shifted the scale of such a state of pluripotency and capacity of variation to that of the undifferentiated cell. The progressive powers of differentiation within a plant or across a species constitute the unfolding of the dynamic vegetal form in growth. Goethe used the term "morphology" in order to mark the study of these changes and of the capacity of change, emphasizing dynamic rather than static (Linnaean) taxonomy.[54] What he points us toward is that there is something very much other than fatalism to be gleaned from the observation of plants. Glory is the capacity of abundance of growth, form, and sensation that is concomitant with the nature of forcefulness in plants; indeed, it is the resonance of its expression, which is always being modified by and passes through the plant but also through its niche, habitat, and the species, which are, in turn, implicated in and respond to it.

One of the characteristic modes in which glory is manifest in plants is through the floral color of pollinated (rather than self-seeding or sporing) plants. Color is dynamic; as Francisco Varela puts it, "Color is a dimension that shows up only in the phylogenetic dialogue between the environment and the history of an active autonomous self that partly defines what counts as an environment."[55] It is something that implies interactions at multiple scales, both within and without the particular organisms and entities engaged in the coupling that produce it. In this set of reverberating interactions, the nervous system is a "synthesizer of regularities."[56] This felicitous term implies the arrangement of structural coupling between cues and receptors; it refers to the work of making the world hang together by working across scales. Such glorious synthesis can be played with and is in turn made in the movement of variation within.

Following such variation, Muriel Wheldale, mentioned above, analyzed the regime of differences in color of the garden flower snapdragons (*Antirrhinum majus*) in order to deduce aspects of the hidden operations of genetic inheritance. The specific color Wheldale studied is that perceived as red and produced by anthocyanin.[57] The production of anthocyanin in plants and in various parts of plants can be due to numerous factors, ranging from the relative exposure to sunlight to the presence of certain chemicals, insect attack, and to the species' particular expression of the chemical in its own fruits, bracts, leaves, sexual organs, and other locations. Woven among these are different inhibitions and productions of other chemicals, giving variation in color and mixes of color. The interplay of colors and the various biochemical, physiological, and genetic expressions of color running variably across species produce a combinatorial explosion of form and coloration, although with certain regularities and variations. In the snapdragon, a series of colors expressed in the flowers runs "magenta, crimson, rose doré, bronze, ivory, yellow and white" and is based around the combination and repression of a range of factors.[58] Indeed, some of these colors may be produced by one or more means, with yellow, for instance, being the result of either a plastid or soluble pigment. In this research, there is a dance of interplay between a then emerging, or revived, mode of Mendelian methodological rigor, made precise in turn by both reason and learned practices of horticultural operation, and an avid engagement with the snapdragon as a hitherto mysterious species, in which the patterns of heritability of flower color were substantially more complex than in most other plants. The production of an experimental practice is a process of inventing an individuation: one that can take reason through necessary detours and encounters, and can learn to arrange it in relation to the synthesizing of realities that are fully or partially occluded from simple immediate perception.

Building on the question of color but also placing it firmly in the question of how to formulate a self, how to compose a grammar of individuation, one of the pathways to follow in reading Joris-Karl Huysmans's novel *Against Nature* is by tracing it as a series of encounters, in which the plant both exemplifies nature and goes beyond it, and in which vegetal glory exhilarates and inspires fear and repulsion. The book can be read as an escalating relay between glory and sensitivity to it. The aristocrat and superlative decadent Jean Des Esseintes is initially a collector of artificial plants: tropical plants reconstructed in Indian rubber and wire, calico and taffeta, paper and silk. Becoming tired with the artificiality of these works of craft, he moves to the collection of roses.

> The roses like the Virginale seemed cut out of varnished cloth or oil-silks; the white ones, like the Albano, appeared to have been cut out of

an ox's transparent pleura, or the diaphanous bladder of a pig. Some, particularly the Madame Mame, imitated zinc and parodied pieces of stamped metal having a hue of emperor green, stained by drops of oil paint and by spots of white and red lead; others like the Bosphorous, gave the illusion of a starched calico in crimson and myrtle green; still others, like the Aurora Borealis, displayed leaves having the color of raw meat, streaked with purple sides, violet fibrils, tumefied leaves from which oozed blue wine and blood.[59]

Des Esseintes delights in the flowers' artificiality, the ability to mimic other substances and incite the passions attendant to them but also to raise them to the scale of parody and analogy, analogously to the way flesh gains color in his boudoir by being arrayed across mirrors and enhanced by rose-tinted satins. Here the plant arranges a sensual abstraction that moves across flesh and metal, worked and unworked materials, the precious and the excremental, the poisonous and the nourishing. Liberation from natural function through generations of breeding and mutation allows for the plants' capacity for glory, the degrees of complexity of its range of possible manifestation, to come forth. The plant leads on and exceeds the refined capacity for dandyism of Des Esseintes, meticulous though he was in his crusade to overtake vulgar reality by an artifice that expressed his individual genius. While in much of the book, the old bag nature is enfeebled and recedes, rendered degenerate by the capacity of artifice of modern culture, surpassed by machines, meager and subordinate to the deliriums of invention, it is occasionally the plant, with its brilliant alienation from the human, that allows the most illustrious ploys of Des Esseintes to come to life in the text. Such is the case in the scene where he orders a tortoise to set off the colors of a new carpet and, finding the contrast between the shell and the tones of the rug too meager, arranges for the reptile's shell to be covered in gold. In order for the gold to reach its maxima of contrast and interplay with the tones of the rug, however, it also needs its own internal difference. The design of a flower is chosen to be encrusted with jewels upon the creature's shell. Most jewels, Des Esseintes finds, are dulled by association with the banality of bourgeois grandiosity: jewels are like traffic lights and sparklers to draw the attention of morons. All of these are described by reference to the everyday reality of towns. These had to be rejected. In selecting the most appropriately powerful hues and intensifiers of light, Des Esseintes's attention is drawn to precious materials that have the hue of various plants—the greens of asparagus, leek, and olive, achieved by chrysoberyls, chrysolites, and olivines. Here, the colors of the plant lie in wait for and excite the excesses of the postnatural and beyond the naturalizations of the mundane expropriations of nature. The material base of these colors proves too much for the tortoise, who succumbs in death to the weight imposed upon it.

The glory of the roses too exults in what surpasses the human, with Des Esseintes's collection being supplemented by roses whose form and color render scars and scaldings, syphilis and leprosy, cankers and ulcers. These wounded and apoplectic glories are more monstrous and gifted with artifice than that can be merely repeated in form by craftsmen. In turn, these magnificent flowers are accompanied by a host of plants brought by traders from the tropics. These forms are fearsome and endlessly obscene in their realization of the chimerical dictum to seek pleasures not yet discovered among new perfumes and stranger flowers.

The narrative has Des Esseintes fall into a reverie, in which he encounters syphilis manifest in the form of a zombified whore figure, "The Flower"—a meditation on the ancient co-development of the human and the virus that has him wracked with febrile pains for days following. Nature is always more than supernature here. Tending to its capacities, the development of the art of perfumery too is one that tracks and interlaces with nature's capacities in conjugation with sensual forces of abstraction. Plants, resins, and flowers exceed their bounds in scents and distillations to be orchestrated and filtered in order to intensify the pleasures and impressions of the nervous system in its flight from the suffocating morasses of the imbecilic everyday.

This capacity for occasionally grotesque glory in flowers is also something that Maurice Maeterlinck's spiritualistic paean to plant intelligence describes as it centers upon the magnificent lizard orchid (*Himantoglossum hircinum*), a tall, multiflowered orchid found in rare patches across the warmer version of Europe, with a concentration in southwest France and in North Africa.[60] Intermittently flowering across the years, it produces a maypole with the delicious aroma of a goat. From each flower dangles an extended lower petal like a ribbon. This twirling streamer is covered, in Maeterlinck's terms, in its inner reaches with "caruncles, buboes and mustaches" and extends out with the bruise purple and greenish-gray lard coloration of a month-old river corpse.[61] Where it is more prosaically described as green, white, and purple, in some of the scientific literature, the analogy to disease is not made at the visual level but in proposing a homology between the spread of plant populations and movements studied by epidemiology.[62] These are abstract dynamics of movement running across distinct conditions.[63] But we can also say, following P. D. Carey's botanical study of the small numbers of the population of this plant, that such abstractions are also highly concrete. He speculates that the number of groups of these plants to be found on golf courses may be largely due to the movement of participants in this particular activity between sites, which allows for the circulation of seeds.[64] The visual grotesqueness of part of the plant is simply a subset of its capacity for association. The flower is a factory for metonyms and relations, a machine for extracting impressions from sensoriums across species but also for drawing out associations between ideas, concepts, and formulae and in doing so perhaps changing them according to their specific conditions and composition.

One such change is the way in which vegetal glory composes a relation to Eros. Michel Serres notes the use of the word "orgasm" by Hippocrates to mark the vitality of sap-filled vigor in an organism and by Jean-Baptiste Lamarck to convey the "physiological tension of sensitive life."[65] This physiological tension is a variant of the arabesque. Here, the ethico-aesthetic dimension of plants joins their reverberations with those of myriad others, a cacophony and tumult running from a tangled bank of plants to a tangled world of beings.

For Maeterlinck, the flower is cast as a "dazzling tabernacle" for the reproductive organs, a term he uses a few times.[66] Maybe there is something in this: Ponge calls the simple brown interior of a dried fig an altar, a chapel.[67] The complex fertilization processes of the ficus indeed suggest a ritual dimension to the plant.[68] A miniature architecture of Eros in a fig implies something that is echoed in sacramental figurations of space. When another media, photography, started to engage with plants—for example in the dramatically close-up images of Karl Blossfeldt, who brought modernist techniques of cropping and objectivity, breaking into the reserve of classical botanical images—the physiology of the vegetal organism widened its field of allure.[69] Indeed, we can say that as photography liberated painting from representation more broadly, and as it, in turn, cut into reality in new ways, canvases responded to plants with particular gusto: from Dorothea Tanning's tumultuous sunflowers to Georgia O'Keefe's simultaneously fleshy and ethereal flowers saturating the eye. The ethico-aesthetic modes of flowers feed into cultures and find amazing allies in certain corners or trajectories of them, in whom their tendencies and characteristics can be kaleidoscopically exemplified. A scientific literature that suggests an understanding of plants as highly sophisticated arabesques composed in-between multiple capacities and forces invites in turn a reciprocation, one to be teased out, in part, in expanded and experimental approaches to the question of living.

Home

In this chapter, our interest is triggered by a figure of a home. It is a house that is active in many Soviet films—nearly all of those completed by Andrei Tarkovsky have one or a few (*Solaris, Mirror, Stalker, Nostalgia,* and *The Sacrifice* are most notable for their homes). The main protagonist does not ever live there, except perhaps temporarily as a child. The house is often one that the parent inhabits: a parent living their life through. In the 1977 film *On Thursday, and Never Again,* directed by Anatoly Efros, a scampish son visits his mother and stepfather in their residence among vast fields and forests.[1] The stepfather is a museum-keeper, and they live on the edge of a national park that they keep an eye on, too. Fields and forest fill the screen throughout the film. The mother dies of a sudden heart attack by the end of the day on which the film is set. She dies in tall green grass at the bank of the river and is buried there, under a willow tree, in the final scenes.

The house of *On Thursday, and Never Again* is not a home for one's usual living, especially not from the point of view of the protagonist, but is on the outside of certain forms of life. It is not for energetic Western pensioners trotting the globe or for debilitating old age but for sustaining other worlds, for constructing new kinds of living. This house is linked to the initiation house mythologized in folktales: there one goes away to get initiated into adult life; here one "goes away" to be initiated into the afterlife. In *Abécédaire,* Deleuze talks about the territory of death. He says that only animals know how to die well and humans die best as animals.[2] The animal is searching for a territory to die in: a gaping hole. The home depicted in Tarkovsky's and Efros's films, which are discussed further below, was another territory to die in—or to figure out how to live differently.

Such a home is full of life, of course: it is a house with more space for new kinds of growing, for attunement, for change. In Tarkovsky's films, the house is almost empty and is therefore full of space. For the child protagonist in the frame or the subject looking at the screen, sparseness and porosity define the house. It is not filled to the brim but instead is brimming with territories ready to vegetate. This house is always in a forest or by the forest and here is where its powers come from. In this chapter we will draw various axes of figuration of the forest and of the home. These axes are not aimed at providing a definitive latitude and longitude of an encompassed territory but rather trace tensions and unfoldings in the formation of ethico-aesthetic compositions such as lives.

FOREST

The philosopher Vladimir Bibikhin offers a phantasmagoric and ecological reading of forest as matter.[3] Bibikhin practiced philosophy by giving a new lecture course annually in Lomonosov Moscow State University. Many of these courses have since been published postmortem. In lectures given in the academic year 1997–98 and later released as *Forest*, Bibikhin reflects upon the first meaning of "hyle"—the Latin word for "matter"—and the Greek word it derives from, "wood." This wood is pure energy, originary stuff for the pre-Socratics. The fire burning the wood warms humans, as does splitting (burning) the atom. The energy of dead wood in a bonfire and of the entire dead ecology of prehistoric forests rotted down into oil, together with atomic energy, all converge in the forest. Wood or forest here is the energy of burning matter and an energetics of matter. Furthermore, vegetation covers the entirety of the human body, linking humans to animals and settling us firmly with plants in the forest. A head's mop of hair, twisted wires of thoughts, bring the human into the forest and bring the forest into the city on furry bodies. Wood is thus matter, in its philosophical sense, something of which everything is made, a historical ecology of evolution, the cosmic forest of thought and curly vegetation between limbs, and the real forest outside cities. This formulation of matter extends from the most physical to the most abstract of its kinds and scales. For Bibikhin, forest includes the fulsomeness of biology, embedding the human historically in the animal kingdom and amid complex ecologies, as well as those of power, poetics, will, and care. While we are used to perceiving the forest either economically or aesthetically, forest, maintains Bibikhin, is not a metric, geographical space. A Russian proverb says that it is possible to get lost in three pines. Forest is a narcotic and intimate space, turning the inside out and bringing the outside in. This is an image of chemical, vegetal, animal, geological, and metaphysical life that is decidedly not that of the passive sort typical of its reduced place in Western philosophy but rather one that abstracts to rethink relations between beings and processes more widely.[4] Forest is thus an originating stuff, ongoing geological formation, a range of species, processes of vegetation, evolution, and ecology, and has cosmological, biological, sociopolitical, economic, and aesthetic histories. As a living environment, it is also constantly unfolding along all these lines. As part of an unsparing industrial machine, the forest is also dying in an unfolding devastation. Bibikhin writes:

> The forest approached them [humans] so closely that it constituted their very skin, their very bodies. . . . The wood, matter, from which everything arose, is akin not to the timber of the carpenter but to passion, genus, the grove of Aphrodite, the poison of cocaine. . . . The contemporary . . . human does not let the forest come close. . . . I could say that humanity has not sorted out its relation to the forest, and a quarrel begins.[5]

One way of answering Bibikhin's call is to find ways of making ourselves at home in such forest, to create a figuration of home sited in this place. In order to do this, we need to rethink the question of home. It means revisiting and displacing the accreted legacy of thinking the origin, beginning, and belonging. The chapter starts with this task, rethinking and affirming home outside the sediments of nationalism, essentialism, purity, and anthropocentric exclusivity. Following on from that, the chapter examines Deleuze and Guattari's figurations of home as territory and the quest for a cosmic home. These leave us with a perfectly working alternative, a quality that differentiates it from those that concern this book. Aspects of migrant unhomeliness and the question of being at home in literature, as well as homes recalled by work or anguished in care, follow on from this. The entanglements of ecology and economy bring about a discussion of the forest as a legally conceived space of freedom, in relation to land cultivation, dispossession, serfdom, and the politics of planting. But to cast shade on possible romanticization of the forest, the chapter also draws on the history of the forest-based trade in furs as constitutive of the state formation of Muscovy (an instantiation of fur-trade-led European colonial expansion of a kind that would also spread into North America). Each of these facets of the forest are rendered here as axes, which may be able to elicit certain kinds of adequacy to the complexity of the forest and of home.

In what follows, state interest, governmentality, and the biopolitics of survival present one of the axes along which the home is stretched. The other axes include idealist–cosmic, political–poetic, and economic–ecological. Here, we track figurations of home as a site, a space, and an ecology, a set of shifting multiscalar compositions spreading across various registers, as they change consistency and reassemble. The forms of composition of such figurations range from existential to economic, from ecological and biological to solemn and sacral: their varying ethico-aesthetic dimensions tie together killing and nurturing, moving away and digging downward. A molecule and literature, passport and cradle, hearth and language, taiga and desert, song and spoor, soil and water, burrow and log, field and frying pan—imaginary and painfully real, homes can be constructed in anguish; they mutate in devastations and are gambled with in the scenarios of irresolvability. The luck or fate of a home is often linked to the politics of planting. This chapter continues mapping the state of stretched suspension that characterizes "the neither here nor there" condition that is often politically mobilized or valorized with woeful consequences. Nevertheless, in this chapter we would also like to create a figuration of home to upend such mobilizations and think through a variety of homes as they change and recompose in the forest of matter.

THE GENUS VORTEX

Today, what does it mean to be at home? A home sustains one's life. It is a place of safety, which affords self-regeneration, both physical (to sleep, to eat, to wash oneself) and spiritual (to rest, to come to oneself). As a productive space, originating and nurturing humans, the idea of home has long been recruited into discourses of nationalism, patriarchy, racism, and other—occasionally more nuanced—forms of repression. In an act of escape from the requirement to declare attachment, or perhaps as an affirmation of nonbelonging, Vladimir Nabokov, who left Russia at the age of nineteen and never owned a house since, famously declared that his home (his motherland) is Russian literature.[6] If one's home is one's native language's literature, does it threaten other people's homes when they hear one speak a language foreign to them?[7] To be at home as a single mother, to construct a home in the child as a migrant moving from flat to flat—does it destabilize those homes built on the models of the nuclear family, national belonging, colonial expectations, and financial fecundity? Further, what does it mean to make a home on an incandescing planet? Can we make homes without destroying what has been there before and without annihilating space for others? Is home bound to the notion of a species?

Home, as a term, has much sociopolitical and existential entailment but is coupled into more than one pairing. In the Russian language, there is only one word that means both home and house: "dom," a derivative from the Latin "domus." The term "home" implies the house as private, something that has been separated, cut off, and sequestered, as Bibikhin proposes in another lecture course, on ownership. Home merges two meanings: something torn off from the commune, from people, and yet full of hope that it will "become alive, take root, become full in itself, independent." "On the one hand, separated, chopped off—on the other hand, whole, by itself."[8] Bibikhin writes about this becoming by itself as apprehending and owning, where ownership itself is double, both juridical (ownership of property) and truthful (becoming itself, of one self). Here, we can see the roots of classical characterizations of home: as an enclosed, cut-off, protected space of safety (juridical both in terms of security of tenure and privacy protection) and a site of forging of identity (authenticity, both personal and social).

In this lecture course, titled *Property (Ownership): The Philosophy of One's Own,* Bibikhin offers a reading of ancient Greek and some European and Russian philosophy from the point of view of what constitutes "one's own." In Hegel, for Bibikhin, "genus" (another term for idea) is both empty and powerful at first, like a vacuum.[9] An individual can only be filled with energy through this power of genus that nevertheless remains cosmically empty, while the newborn individual is blind to who they are and what kind of power gave birth to them. Both the individual and the genus itself are born a second time as an idea, finding their ownness. It is a vertiginous circular movement of repeated mutual birthing to find itself, one's own. "This own sucks into itself first the family, then civil society, then the state, and the world not as an awkward sum of states, but as the highest absolute truth of the world spirit. . . . Only in the world does the own find itself."[10] Hegel here, for Bibikhin, is the most obstinate thinker who requires the most obstinate reader. Bibikhin invents a method to read Hegel, whose dialectics, he declares, can only be understood in relation to the fight for freedom of thought, and is a relentless confrontation staged by Hegel against everything that is not one's "own."[11] Inspecting everything as being a priori provided, being enmeshed in the *not-own*, Hegel, a philosopher of law, is relentless in seeking this almost ungraspable apprehension of the own. It is only the own of will, the world spirit, that, once reached through the spinning centrifuge described above, can return things to themselves, so they can truly own themselves. Only the will can be free and is a "breakthrough to the idea, i.e. genus, i.e. properly own, that bestows the thing its own self, endowing it with *ap-propriating* as a return to its own essence."[12]

The detailed working-through of this circular nature of ownership (as property, enshrined in law) and the ownness (of looking for oneself) in ways in which it bulbs out to encompass other people (the social, what could now be called maintaining a sociopolitical identity) and the national, worldly (now the questions of national identity, migration, globalization), all returning to the thirst for oneself, is most striking in the ultimately state-centric Hegelian system. In Hegel, the bottomless whirlpool of seeking the own of oneself becomes more open to the worldly, global whole the deeper it slides into self-absorption.[13] For Bibikhin, nationalism cannot be grounded in a thinker of such obstinacy as Hegel: instead nationalism can only operate, piling mistake upon mistake, via reductive schemata.[14] There is nothing to sustain nationalist or fascist projects of defined purity when drowning in the Hegelian vortex of fierce seeking for one's own. Nationalism, says Bibikhin, will not have the slightest idea that the general, the worldly, the universal, the for-everyone is head over heels in this seeking of oneself and one cannot find one's own by rejecting the universal, the common, the worldly, because one's own is never available solely in one's own. It can only be born, generated through the genus, through the idea, that can never be reached until everything, the world, is dialectically encompassed.[15] Hegelian idealism provides a strong thread of thinking home. It is one of a striving search for oneself that could return you to you, or give birth to you, by encompassing different aggregates and relations, of loved ones, of communities, and of the world itself. There could be no essence here that could ground nationalism, only the vertiginous seeking, with a hole in its heart, an emptiness in its center, for which no plug is available and that can never be filled in.

Bibikhin also turns his attention to Heidegger, whom he translated into Russian. For Heidegger, he claims, to think dwelling means thinking the place where "the truth of Being happens."[16] To dwell, to build, "to remain at peace with the free," to keep, to spare—all are etymologically sourced meanings of a starting point of "building on earth" (a shelter, an abode, or a home), which finds itself at a counterpoint to but also houses, sustains, and unfolds the entire Heideggerian system.[17] Dwelling is thus the fundamental character of Being. In Heideggerian ontology, Being is always incomplete, so that the prospect and actuality of death occupy the main role in its emergence. Being, oriented toward mortality, is unable to be its own foundation, hence Heidegger's call for the dignity of man.[18] The vertiginous incompleteness of the own of Being is initiated into and practiced in the dwelling, whereas everything is sustained and emerges out of a hole of being looking for itself.

Without doubt, Bibikhin offers a generous rereading of Hegel and Heidegger. Dedicated to a project of true philosophical thinking that, in his view, can never be of "use" or be "wrong," he is sympathetic and humorous. Indeed, both of these two make great straight men to the funny guys: writing on Hegel and humor, Bertolt Brecht noted that Hegel "has the makings of one of the greatest humorists among the philosophers. He had such a sense of humor that he couldn't think, for example, of order without disorder. It was clear to him that directly in the vicinity of the greatest order the greatest disorder resides, he went so far that he even said: at one and the same place! . . . His concepts have always been rocking on a chair, which at first makes a particularly comfortable impression, until it falls over."[19] The foundations of the universe are ultimately an excuse for slapstick, and the means by which it must be carried out.

Notwithstanding the ontological grounds for humor, the histories of the "use" of philosophical furniture comes strongly into discussion with the emergence of what Rosi Braidotti calls the "studies areas": feminist, postcolonial, cultural and media, critical race, animal, and plant studies among them.[20] Each of these responds to particular sets of interlinked crises, and as such they are replete with formations that cross the threshold of home. Scholars studying indigenous dispossession and especially the interrelations between race and ownership widely document the ways in which settler colonialism derived its juridical property relations from the philosophy of the "own" and the ownership inspired by German idealism and Lockean empiricism.[21] Similarly, Étienne Balibar's *Identity and Difference: John Locke and the Invention of Consciousness* links the ownership of property with a self-possessing subject, echoing Bibikhin's reading of the relationship between ownership and the own, through personal identity and labor in terms of property and propriety.[22] Here, every man is a proprietor of his "own" things, such as, for instance, thoughts and experiences, which are appropriated (to our selves by consciousness) to serve as personal identity. There is a pragmatic relationship to identity as an own property that is extended to labor and juridical responsibility in terms of ownership. Brenna Bhandar, among others, further sets out the mutually assured relations between property and propriety as a philosophy of colonial governance, a foundation for slave ownership and the colonial appropriation of land.[23] The figure of the self-possessive individual necessarily includes the possession of others: of nonindividuals. The constitution of one's "own" is through a circular movement: to cultivate, to grow crops, to feed the family, to sustain the nation, to control the world. Here, John Locke, German idealism, literary sentimentalism, and Romanticism were efficiently put to use to build the project of the nation-state and is a continuous source of inspiration for some state doctrines: for instance, those of modern political Zionism, based on attachment to land and agricultural colonization as a tool of Palestinian dispossession. We will return to this discussion later on. To conclude this philosophical interlude, it is worth noting that Bibikhin's work on property should be read in the context of the history of the Soviet Union. Alexei Yurchak named this time well: *Everything Was Forever, Until It Was No More*.[24] Bibikhin gives his lecture course in 1993–94. After nearly eighty years of no private property, the "until" has arrived and Russian society sees the rapid privatization of public resources: infrastructures, factories, land, and buildings. This lecture course is sustained by curiosity about what it can possibly mean to go back to a regime of private ownership. What is this thing, property? Can it be any good? In such a context, Bibikhin's conceptualization of forest can never be thought as private property, even if it belongs to itself.

Looking for oneself, asserts Bibikhin, will not bring one to the essence of oneself-in-waiting, and practicing the capacity of being, at home, will not provide one with any authenticity of self. There is a hole in the center of being that prevents it from holism. In Bibikhin's reading, the Hegel-Heidegger pole depicts home as a space for being "in itself," that is, a longing for something essential, absolutely core, and yet something that can never be reached, found, or fulfilled. For Bibikhin, this is a productive seeking. But it is also a firmly anthropocentric system, ultimately a model for the anthropocentric homeland, and, as we have seen, it can be a colonial and racist home at that. This homeland is for *some* humans, and there is a hole inside.

NATAL-COSMIC

It is a bit obvious to set Bibikhin's reading of Hegel and Heidegger against Deleuze and Guattari, given that Hegel was the philosopher Deleuze resented for establishing negation as the exclusive mechanism of difference. While Deleuze considered Heidegger's method to have opened philosophy up to the questions of being alive, and in broader terms to have articulated certain facets of the question of being as a problem and as a project, it is a morbid and pompous vision, no matter the prolixity and nuance of its apparatus.[25] Yet, for the theme of home, Deleuze and Guattari's posing of the question of a place of living, which is also a question of what living does, can do, and the play of finitude and infinity in variation that manifests in the problem of how to live, has been used to build one of the few alternatives to the quest for the absent core. Missing something essential that can never be fulfilled has leaked into thinking habits, images, metaphors, and kitchen philosophies in their idiosyncratic and syncretic intertwinements, pop invariants, and at times horrific applications. One can see its traces in the images of lonely jet-setters and nostalgic Russian millionaires, the rhetoric of the tourist industry, the angst of the many kinds of paupers of London, in stereotypes of Indian doctors in the UK or of programmers in Silicon Valley, in the shaky victories of Brexit and of Donald Trump, and many other cases where an unfillable void is either politically expedient or fulfills the needs for a powerful fantasy. The siren calls of imaginary homes, always inevitably false but nevertheless painful, are maelstroms, for both those coded as settled locals and nomadic cosmopolitans. Reframing the search for home ecologically, away from the simplicity of the vortex, into the chaosmosis of composition—into the ethico-aesthetic dynamics of plants, animals, relations, and movements—is what Deleuze and Guattari offer in *A Thousand Plateaus*.

In the well-cited chapter "Of the Refrain," home is first of all the making of territory by an animal. Something (bird song in their first example) starts a home, organizes space around a center—the song. This center produces space, inflating the abode or home as an interior zone. The center, the mark left by an animal, is linked to genus; it carries with it the forces of the earth, the natal.[26] And yet this natal is ambiguous, projecting the "intense center, which is like unknown homeland," into the outside, linking it all the way to the shifting forces of necessity and chance that make the cosmos.[27] Such a home is thus a foundation that can become creative, not acting as a force pulling us down into the void—the black hole of ancestors, death, and closure—but opening up into the universe, linking the earth vector into cosmic forces at many scales.[28] Here, "the problem is no longer that of the beginning."[29] Deleuze and Guattari liberate home from the thought structure that links dwelling to the deep past of origin. The fire that creates the inside in the outside, the domain "beyond man's earliest memory," as a lineage that speculating on home inevitably invokes is here opened outwardly.[30] The natal, the own of home, here is also unreachable, and the movement it inspires and compels is to go outside, but in a radically different fashion. There is no hole in the heart of being, only endless differentiation of a monistic whole. In Deleuze and Guattari, the impulse to the outside of the territory and into the cosmic is posthuman and not solely biological: it is not about the human, nor even more generally about the animal, but rather identifies the forces and intensities that each of their scenarios individuates and sets in play, and it is not about the world but the fructiferous and annihilating cosmos.

It is fascinating how profoundly linked the question of home, dwelling in Heidegger, territory, and deterritorialization in Deleuze and Guattari, is to the main questions of their thinking. As soon as the question of home is posed, a few paragraphs follow that act as succinct expressions of their philosophical systems.[31] It is like a pop-up book. The natal for Deleuze and Guattari is ambiguous, at times to be found outside as belonging to the territory tending to deterritorialize, and in absolute deterritorialization, to become absolute, cosmic. This absolute, cosmic natal is everywhere and in everything, like Spinoza's God or the cognitive ocean in Tarkovsky's *Solaris*.[32] Cosmic philosophy, testify Deleuze and Guattari, thus articulates the forces of the cosmos directly, not as Hegelian will (for the world and some humans) but through the technique of exploration of direct and convoluted relations: material forces. And as their program, they write: "The forces to be captured are no longer those of the earth, which still constitute a great expressive Form, but the forces of immaterial, nonformal, and energetic Cosmos. . . . The essential thing is no longer forms and matter or themes, but forces, densities, intensities."[33]

In the many hearts of chaos, joining with the cosmic forces of the future, is the earthly space of home, both innate and acquired, that territorializes into the social, cultural, collective, and that can give access or rather leap into the cosmic and its cadences of energies and variations, posing a program for philosophy: to cosmically think unthinkable questions. The aim for the thinkers of home in this fashion is thus to make the homeland into a plenitude of cosmic homes.

Deleuze and Guattari offer a highly affirmative reading of migration: "One leaves all assemblages behind, . . . one exceeds the capacities of any possible assemblage, entering another plane. . . . These are no longer territorialised forces bundled together as forces of the earth; they are the liberated or regained forces of a deterritorialised Cosmos. In migration, the sun is no longer the terrestrial sun reigning over a territory, even an aerial one: it is the celestial sun of the Cosmos, as in the two Jerusalems, the Apocalypse."[34] Migrations bring about more than two suns. Nightingales fly to Russia, Northern Europe, and Southwest Asia in order to sing their beautiful songs there in May and early June. For the rest of the year, having raised their young silently and flown back, they live unnoticeable lives as little songless gray birds of sub-Saharan Africa. They sing under the gentle sun of the central Russian countryside or in woods designated for house-building, and they are silent under the sun of Africa.[35] They fly over half of the globe, following celestial movement, until one sun becomes the other sun, the nonembedded sun, a cosmic sun under which they live more than one life.

We wish the flights of people could be similar. Leaving the territory, deterritorializing, no longer existing with reference to the one-alone of German Romanticism (where it arises as a legendary and solitary hero of the earth) or the one-people of, in the Deleuzoguattarian figuration, Latin and Slavic cultures ("the hero is a hero of the people"[36]), the expressive seeking of home is neither the existence of the earth through one-alone nor the existence of the people through one. Expressed very differently to Bibikhin but covering the same territory regarding nationalism and identitarian belonging, this argument is for new openings for these forces—into the cosmic rather than simply earthly empowerment. In *Nomadic Subjects,* Rosi Braidotti relates her story of movement between countries and languages as a child and as an adult, as a liberation, a gift of global awareness, a truly internationalist freedom, and a foundation for her work on a multiple, unbound subjectivity.[37] What kinds of home might such a multiple subject develop, to be produced as such? What are homes and how are they assembled, and how do they decay in those troublesome places, where one finds oneself intermittently heading for the cosmic or looking for oneself, and becoming variously stuck and unstuck?

One aspect, then, of the vector of home is Deleuze and Guattari's natal that deterritorializes into the cosmic in all its hopefulness. The other that asymmetrically pairs with it and contests it is what Bibikhin formulates as the "own," that, through Hegel and Heidegger, obstinately spirals into the hole of genus. Starting with and through home, the intimate expands into the spectacular, the subjective into the grandiose, the own into everyone's, individuated into the absolute, germinating into differentiated, bodily into artificial, biological into aesthetic, simple into complex, past into future, and earthly into cosmic. Home contracts into time immemorial or seeds of primal energy and expands to sustain differentiation, act as reserve or a blueprint and transformation, encompassing vast and changing spaces that may sprout from the minuscule. Vitreous home, sustaining vision—how often it goes bad! What kind of vitreous degeneration happens to it? Which floaters populate it? The condensation of fibers producing entertaining and elusive shapes and forms in the line of sight, the sticky opacity, are not simply obstacles on the road to a better society, life, or world but the bleak undertones, if not devastations, of the events of life—of long-standing and newly emerging struggles—of vital processes that just occur and may dissipate but not disappear. So much lingers at home that usually enjoys an affirmative entrance into speculative discussion that it is tempting to say, even just out of spite, that there are more deformities, bastards, and evolutionary dead ends carried through the home than well-ordered formations, sustenance machines, or beautiful unfoldings. Yet we are not on a search and negate mission—the dialectic of resentful sufferers, of unhappy consciousness—nor on a nihilistic one; nor is it possible to side with the fitness instructors of pluri-positivity. Home is an ethico-aesthetic process, one that could certainly become a hole, and that may recognize its workings in terms of the cosmos but is not singularly sought out by the humans. Even when we choose an anthropocentric view, home is where agency and subjectivity can be radically altered or removed. One can be as much kicked out of one's home by becoming a parent as out of one's body by becoming a cancer patient; and this is an event in difference, not one mapped by a logic of negation. The cosmic is inspiring but can be rather indifferent to the clattering human, biological, and ecological occurrences. The inhuman cosmos tantalizes in a different way than an imaginary genus hole in the center of oneself, but finding an adequate relationship to it may be difficult to achieve—and that is also tantalizing. As this chapter continues, it looks for the sight of the home and aims to discern some specific experiences, sensibilities, and dramas that occur within its fluctuating parameters in today's world. This chapter is interested in the various migrations, in getting stuck between the worlds, which happen on different axes. The stickiness of such things indicates that these spaces are not enveloped or instituted by black holes, but neither are they always fully, or directly, cosmos bound. The condition of in-betweenness, undecidability, and irresolvability characterizes this first worldly cosmic, human-animal axis: home is gathered on these lines, neither here nor there, neither unachievable nor fully realized, composed of multiples of notions that are not dialectic although they are sometimes paired. Spread out, it indicates a variant mode of the organization of home: a longing but for nowhere in particular. This home longing is an expressive, aesthetic converter that is constituted by its always available cosmic malfunction.

MIGRANT

The home that offers both protection and authenticity—how does it work? When one has a flat in a large block, built on land that is leased, on, say, the eleventh floor, a home is effectively merely some air between plasterboard walls. Despite the miracle of such a thing, homes are better understood by their symbolic acts of delineation, supported by social and juridical norms. As a space that separates, it also, by the same action, conjoins, pulls together. Hence the home is a space of regeneration not only of a self but of social collective selves, cultures, geographies. The home, as mentioned above, is often discussed as a site of both individual authenticity and social belonging simultaneously, even more so if one has to individuate in conflict.

To talk about the migratory condition, often one undergone together with destitution, is to talk about what is changed in the condition of the genesis of the individual that always happens in relation to the collective and to the outside. In Homi K. Bhabha's figure of the unhomely, the illusory and symbolic home in the new land falters because even if one has a house, the social, the political, and the collective it separates from and adjoins to is alien or can be found to be repressive. Bhabha writes, "The unhomely moment relates to the traumatic ambivalences of a personal, psychic history," and he further records that such histories are marked in certain ways: "By making visible . . . the unhomely moment in civil society, feminism specifies the patriarchal, gendered nature of civil society and disturbs the symmetry of private and public."[38] Here, the alliances of home come together under a different light: whose law, protection, privacy, and authenticity and what kinds of traumas?[39]

The Hungarian Swiss writer Ágota Kristóf expresses the difficulty of even describing her condition of loss.

> How can I explain to him, without hurting his feeling, and with the few French words I know, that his beautiful country is a desert for us, the refugees, a desert we must cross in order to arrive at what is called "integration," "assimilation." At that time, I don't yet know that some of us will never arrive. Two of our number returned to Hungary despite the prison sentence that awaited them. Two others, young bachelors, went

further, to the United States, to Canada. Four others went even further, as far as one can go, beyond the great boundary. These four people of my acquaintance killed themselves during the first two years of our exile. One with barbiturates, one with gas, and two others with rope. The youngest was eighteen. Her name was Giséle.[40]

Moving as a child or as an adult; being children of immigrants, second-generation immigrants; being forced to move; fleeing or choosing to, with the latter inciting guilt and the former atrocious trauma; forcing the narrative into a happy end, with ultimate success as a reward for persevering—all these specificities constitute endlessly differentiating migrations. Kristóf, already a writer, though only twenty-one years old, fled Hungary with her husband and a four-month-old baby into the French-speaking part of Switzerland after the 1956 invasion. She worked at a factory for the first five years, and it took her twelve years before she could write in French. This muteness of a writer, the scarcity of her tongue's means, is her "unhomeliness," lurking on the peripheries of the home continuum.[41] While the "unhomely" still implies a home—not as absence, orphancy, or an outside—it offers something composed of constantly diverging, faltering homes that can rend and rack those that they happen to.

What Kristóf describes as her life story is a situation of getting stuck in-between the inwardness of roots-seeking, sustaining some people, and the cosmic forces of the rhizomes handling earth. One finds oneself hanging over the cliffs of nostalgia, the thorns of the native language, the abyss of the homeland, under the vinegary blue sky of the world heading for democracy. Constantly feeling unable to express something in the new language, being served words that cannot be of service, not becoming a good cosmic girl: a planet might have the same problem. Imagine indeed, as does Mark Fisher, the enormous difficulty of being something as full as a planet, like Tarkovsky's *Solaris*, in trying to communicate with humans, having only their defective psyches as receptive antennae.[42] Unfortunately human, too embedded, stuck in time and space, we cannot always cohere with the lucidity and composure of eternal protons joyously making their way through various compositions. W. G. Sebald writes in the *Emigrants*: "And always . . . one was, as the crow flies, about 2,000 km away—but from where?—and day by day, hour by hour, with every beat of the pulse, one lost more and more of one's qualities, became less comprehensible to oneself, increasingly abstract."[43]

The in-betweenness of being in a supplementary or supplanted language, the unhomely moments, relate to Sebald's awayness and one's own increasing abstraction. One that is always two thousand kilometers away; away from where? Just two thousand kilometers away. This *awayness* is not about longing for authenticity plugged into a language or a nation-state; neither is it a state of hybridity, which is of multiple mutating presences. What *can* happen to a migrant is becoming away. Becoming away tends to exclude or inhibit full affirmation of the natal, not being cosmopolitan, and, unfortunately, is also a fair few kilometers away from the cosmic. Awayness is the thinning out of different kinds of belongings. While it can be a position of privilege and luxury, it is not, statistically speaking, necessarily so.

Guattari writes that being a rhizome as a subject does not exclude being tied to certain particles of culture, climate zones, or languages: "Other institutional objects, be they architectural, economic or Cosmic, have an equal right to contribute to the functioning of existential production. . . . Territory or homeland doesn't necessarily involve searching for one's country of birth or a distant country of origin. . . . All sorts of deterritorialised 'nationalities' are conceivable."[44] However, the problem is not of being tied to the origin; the problem is how the existential production of the embodied specimen and their capacity to create is sustained. Language falters and becomes of no use for Kristóf the writer, and although no one or nothing except for a stroke or other brain damage can take her native language from her, the desert she has to cross to arrive to a space of awayness as her only hope is very vast. Neither is a return possible.

Is this a home of a nomad? An away home of becoming increasingly abstract is a home that has anguish. Suchlike and dissimilar homes are widely discussed in the studies of detention camps and refugee centers; in diaspora studies; in trauma and memory studies; and in theories of migration, exile, and nomadism. Understanding diasporic spaces and transnational places has a rich legacy, with some work celebrating the displacement of unitary identity and others attending to the violence of privilege, racism, misogyny, social deprivation, and politics of exclusion.[45] The studies of migration, diaspora, and decolonialism pay detailed attention to the various figurations of home in transnational journeys, problematizing longing and desire, origin and locality, community, struggle, and inventiveness. This home revolves around human-centered empathy, suffering and trauma, on one hand, and the distributed nomadic call to spread out, on the other. As there is no resolution in the character of home, there is no exhilarating catharsis for the migrant. No full humanist fulfillment for the human. No boundless cosmic drift for the nomad. A thinned out, abstracted home. Two thousand kilometers away, or perhaps twenty thousand kilometers away: from Europe, Enlightenment, humanism, the upper limit of the Earth's atmosphere; from folktales; from the abusive commuter; from the imagined origin. This home is a tool to fight hardship with anguish, irresolvability with luck, to root through devastation with diverse forms of intelligence.

POLITICAL-POETIC AXIS

Home as provisory of shelter, reproduction, and empathy in relation to the sustenance of self and others is ambivalent, productive of recreation as well as of violence, repression, and disfigurement. It is not very surprising that the idea of home, suspended on its cosmic way between German idealism and postcolonial and feminist work of fearless care, quivers: it is dangerous and traumatic to be homeless. And yet related kinds of shaking can also happen in something set up as a home. Are these the differences of degree, of quality, of composition, of coordinates? Can one be partly homed, half authentic, almost protected, semiself?

Hannah Arendt's construction of the notion of the household brings us back to Bibikhin's Hegel. In classical Greece and ancient Rome, argues Arendt, a man without a house, a location of his own, does not possess himself and therefore cannot be free.[46] Possessing oneself and liberating oneself from the drudgery of necessity is the prerequisite for any construction of freedom. This freedom could only be political. For Arendt, home was first the site of conflict, violence done to oneself and others through the order of bodily interdependency, and hence not a place for equality and freedom. As the concerns of the home become collectivized, in the current condition of biopolitics (a term Arendt does not use but that partially follows from her discussion), the political realm loses its exclusive ability to forge freedom. Public and private blend into the social. However, we can imagine that for Arendt this does not relocate the production base of freedom into the house.

Gaston Bachelard, from the opposite position, speaks of home as the site of freedom. This freedom is aesthetic, foundational to all other freedoms. His house is a home of poets, children, and dreamers, and is a condition for their ability to daydream. Bachelard's home is poetic. The house here is full of images that extend before the earliest memory, and, therefore, beyond the last one too, a vessel for dreams of "mankind."[47] This home, the childhood home, is a personal intimate relation, a reservoir to go back to. Its secrets can never be known objectively. Its corners are those to which one needs to take a candle. The aerial and the terrestrial, the ground and the underground, cosmos and anthropos meet in this mnemonic aesthetic structure. Here, "dream is more powerful than thought."[48] Life is larger than reality.

The Arendt–Bachelard vector is another antagonistic and asymmetrical pair that makes up home. This pair spreads out along a political–poetic axis. On the one edge is a home of natural necessity, an opposite of freedom, though with its maternal relation to invention that may break this vector, and on another is a home of daydreaming, an enabler of freedom. One is a home that represses; the other is the one that enables. The reason remains the same: the home acts as a space reserved for the genesis of the body in the absence of sociality; what one finds a hindrance is a reserve for the other. It all depends on the kind of action to arise from this freedom, as an end goal as well as the precondition. A household may exist, or a place of residence, but it may not coincide with home: poetic memory is distinct from a social one. Or is it? The political–poetic axis of home begs for feminist and postcolonial questioning. Home is momentary, in its sustenance of needs, and enduring, as the site of the history of relations of humankind. It becomes a web enabling the established violence of the past to sustain itself into the openness of the future and the conditioning of whatever happens to be the current outside to mutilate the long germination of the inside. Arendt–Bachelard set up this axis, but they are both at another tangent to the woman, to the migrant, to the person of color. Bachelard's thinking, daydreaming, belongs to the cellar, to the attic, to the terrace, to the edges and insides of spheres constructed there. What if one does not have any, will never have any of those? Whose knees, hands, and thoughts sustain the cradle of the house of the daydreamer? A home to be constructed, to work on ceaselessly, to be renewed, sustained, and imagined, is a site of colonization and employment. In Arendt, the women of the house, that is, the women of the premodern period and early modernity, are delegated to the domain of the body, the private domain, and therefore excluded from politics of the public space as well as from reason and, logically, from daydreaming. When the individual transforms into a Foucauldian citizen and the private-public transforms into the social, this elasticity of the same space of personal-communal remains and is emphasized. A woman who has to respond to every squeak of a newborn: Can she not ever think, daydream, or make things up? Can she only ever sustain the other's daydreaming? Do the maternal features of the home Bachelard commends—in that it shelters, protects, and allows—suck the woman dry?[49] Cleaning and daydreaming, serving and acquiring a political voice—what kind of home could these thrive in? Daydreaming self-valorizes: how do we reconcile commanding to be clean and commanding to be free, or to be free from commanding?

If thinking belongs to dwelling for Heidegger and Bachelard alike, with the figurations of the Black Forest hut or a French countryside house, what about immigrants, non-Europeans, or refugees? The political–aesthetic, public–private, social–intimate, painful–away axes extend the ways in which home is suspended, undefined, irresolvable. Home, as a Romantic cultural product of the nineteenth century's formatting by the bourgeoisie—and Bachelard's figuration of home can carry such undertones—is displaced in Bhabha's unhomely moment, unhomely houses of postcolonial theory and fiction, that mark "historical displacement," the "half-way house of racial and cultural origins," and "the in-between temporality . . . in-between reality." Bhabha mentions the aesthetic distance of a "difference within."[50] This is not Bachelard's solitude but an in-betweenness, and muteness, which centers the dilemma of various kinds of work—on self and others; on society and poetics; on history, politics, freedom, and dreams—on work at home.

MOTHERHOOD/PARENTHOOD

One can be alienated from a home as a woman, a parent—violently, when there is no space for them anymore as a thinking, imaginative, daydreaming being—or a carer, drying out on a forever switched-on frying pan of needs. One can be displaced, rendered irrelevant by her sociobiological function or by an accident or disease. It is not untrue that the language of violence is of private pain. Instead of opening up, vitality can produce closures, black holes, and the lines of impoverishment.

As much as becoming a mother is about plunging into utter vulnerability, or weakness, it also appears in itself to be an irresolvable requirement of presenting in two places at once: needing to care for the child and "being there" as well as "being elsewhere," working, earning—to care for the child. These two mutually exclusive though circulatorily connected things imply either dependency or poverty, and often both. As much as parenthood is sappily celebrated as an event of love, it also happens, statistically, to be the single event that pushes even previously affluent parents below the poverty line. Home in this condition is as much a place of comfort, care, and safety as struggle, neglect, and disappearance. While the house is usually described as a place of belonging, it is also a point of departing, a passage of leaving.

Desiccation on the frying pan is not about postpartum depression or the difficulty of parting with the solipsistic habits of youth. As with other states explored in this book, such anguish of motherhood/parenthood does not have a clear causality, form of resolution, or end state. Certainly, parenthood is a social affair and society's programs and paradigms may assist or inhibit its unfolding. Adrienne Rich, Jacqueline Rose, and many others have written against essentializing mothering bliss.[51] A home can be an instrument of exploitation, sustained by the ideology of patriarchy, colonial legacies, and other similar "gifts." Thinking through and challenging societal joys is not a fight to reclaim the bliss of parenting against the repression of the value-extracting machine. If only it were that simple. Parenthood is an axis balancing love and madness, dedication and suffocation, giving and taking, sacrificing and requiring sacrifice, soothing and scaring, indulging and mutilating, disempowering by protecting. Even gray horizon, extending infinitely. Deep anguish. Dark but quiet. Inaudible and glacial. Opening up, slowly, like a gray flower bud.

The black hole of parenthood can make home into a black hole. Parents may act as black holes whereupon endless giving becomes endless taking. The gift constructs eternal indebtedness. We are the children, owing an eternal debt for the maternal life we took. We are the mothers, sucked dry and passing the debt on, for generations. We are the children used as weapons, arms to kill oneself with, hostages in the negotiation of existential meaning, instruments of parental wars. We are the parents conducting military operations, winning and losing, in pain.

Isabelle Stengers talks about "the feeling of irreversible and catastrophic loss [that] offers affinities with feminist thought, which attempts to weave together thinking, imagining, and practically enacting."[52] Such catastrophic loss—how does it relate to not only human anguish, to being attentive to and making home in anguish as a species, as a migrant, as a mother? When you have a child, your home disappears. The new home in its place is a process without a place, a changing process around the changing child. The process mutates; you adapt to it. You are waiting until you can play a lesser role in the process. Then you can live differently; your home will be where there is no process, a home where the process is thinned out. And where one is not a carer. Not the central subject. Not the human.

We need more skepticism about humans—mothers, fathers, and children included. Do critical studies have to have a healing function? The selfishness and sacrifice (of parenting) are constitutive of self-centeredness and self-importance. Judeo-Christian admiration of sacrificial motherhood enforced by Renaissance-based exploitative adoration of womanhood or humanity in parenthood is a trap, concealing human moral superiority always ready to go on its destroying mission. Instead, we affirm ambivalence. Some anguish can and needs to be persistent and permanent. Pain does not produce worth. Suffering can be pointless: an instrument people use to abuse others while simultaneously suffering themselves. Someone can feel and be disempowered while being empowered in other ways, generating monstrous consequences for their dependents. In politics and the economy of pain, there are trade-offs, loans, banking; in its rolls of the dice, there is lucky plunging and fateful folding.

In the figure of the home we are trying to cook up here, one is off-center; there is no center. The home is not organized around anything. There might well be pain, but it is not productive of the home itself. There is nothing specific one must do or experience there. No redeeming function. A posthuman home, sustaining the dreamer without requiring a servant. Sustaining a carer without requiring a sacrifice.

Such a figure of home is not a home of the mother, a process of a home. Not an immediately or uncomplicatedly natal home, nor certainly the origin home of the genus. Not a hearth. Not a home of the migrant, full of pain. Not a home of the carer, filled with duties. Not a home of a nationalist, empty inside. Not a people's home. Not a territory, under a starry sky. Not a bunch of things erecting a place of inhabitation. Not an architectural sustenance of the memory of humankind. Not entirely a private space, nor a public space. Not only unhomely. Not only a dreamer's home.

OUTWARDNESS

The home in the films mentioned at the beginning of this chapter is not a space to sustain the household with its hierarchies, types of power, and needs. In *Mirror* (1974), the house by the forest where the mother and son relocate for the war years alternates with a flat in the city in which the boy is growing up and which is no less full of daydreams.[53] The house by the forest was only temporary and now exists ephemerally. In *On Thursday, and Never Again*, the house is where the mother moved to live with her new husband at an undisclosed moment in the relatively recent past. It is again not the house where the main protagonist was born or grew up, and it is not where he keeps the paraphernalia of his early years.

There is an ephemerality to such a house; it is certainly not where one's "roots," juridically speaking, are. It will not show on one's birth certificate. This home is always temporary, and it does not make a claim of origin, pinning one down to the inward drilling of a house's foundations or to staking a claim on land, to property, one's "proper self," rooted in the accident of the place of birth. Neither is it a place of the rural household that grows its own food, nurses its babies, and buries its dead there, sustaining it as a locus of their genus, a cradle that links the churning of soil, whether for food growing or for digging the graves, to sustaining the family, the nation-state, ethnicity, or people. This house is not the pillar that sustains the rule of humans on Earth.

Instead, this house is open outwardly: it is unclear to whom it belongs and it is certainly not the property of the people who occupy it. It is worth remembering that the culture these films were created in did not believe in private property. This house is not the base of the household terrorizing its women for production of genetically authentic offspring that can inherit the house and the land. Neither is this a mortgaged house: it just does not exist in the order of property. The occupation of the house is always temporary. If one's roots are there, they are of the order of imagination, perhaps in daydreaming of a Bachelardian ilk or of another kind, rather than blood, genetic, or citizenship ties or ownership rights. This home is open for a slow movement of people and other beings through it, with long encampments—but not too long. The home is waiting, serene in its outwardness. It is fragile and can easily disappear: Tarkovsky burned at least one house per film.

The home is open outwardly materially, spatiotemporally, and aesthetically. As a wooden construction, it appears temporarily among other kinds of wood: trees, logs, poles, stubs, and planks. Filled with the products of the forest, it enlists tables, benches, wooden window frames, as well as living plants and field flowers. Always surrounded by trees, positioned at the forest's edge, the house is next to an unmown field of undisturbed green. The house is extended between the forest and the field both internally and externally: the forest comes in, constitutes the inside, and leads outside. Turning outside in and inside out, this home attains a certain plasticity. It is not the supple subject of its relationship with the forest. The forest— remember Bibikhin's expansive notion here—includes the house all the way down to its wooden floor and fences, bushes, animals, particles, dreams, sounds, movements, and humans. The humans, therefore, are not the subjects of the house either. The house is outward in its forest state, and humans transit through it, as they move through the forest temporally and spatially: drifting through the time of evolution and rattling over the globe economically and ecologically.

The forest surrounding the home is not a garden or a landscape; it is a way to escape habit, the family, anthropos, and its culture. While some of the inhabitants of the home may belong to the anthropos in escapable ways, with such a home they can displace themselves a bit, from the central, subject position. Still able to take responsibility and exercise agency (nurture if needed, protect, take sides), still bound to their bodies and negotiating their needs, still emotional, fully material, and metaphysically important but not the tsars of the universe, not the head steps of cosmic staircases, not the center of the forest. The forest has no center: it changes shape and amasses power in nonlinear ways.

It is perhaps here that the idea of the home for initiation into something of an afterlife finds its nest. It takes "going away" to see a home that is not full of birth and food, ownership and blood, where life is not the conquering, the capturing, the hungry leaping out onto things. "Going away" to figure out how to live differently is a modality of becoming that is acute, outward, tuned into temporality, material transition, and into the matter of the surrounding forest. It is tuned into itself as a forest of Bibikhin's kind: matter, energy, animal and plant, ecologies, change and stasis, nature and culture, and thus in its intermediate state articulates something of a relation to the cosmos. This house is not a grounding point, stable and steady, for sustaining nostalgia and homesickness but a place of moving—a passage. An antimonument, it acts as its own removal process, removing itself as a reference point. A home is not just standing there; it is moving, just slowly—producing space. This home is discharged—it parts, giving way to a different space, with another consistency, new inner resistance, new forms of perception, and new ways of living.

FOREST-CULTIVATION

A home in Tarkovsky's *Solaris* and *Mirror* and in Efros's *On Thursday, and Never Again* but also in some films shot twenty years later—for example, *Mother and Son* (1997) by Alexander Sokurov—is always built into a forest.[54] It might not be a dense wood but a forest edge, rolling into a meadow and dissipating into freestanding trees, scattered willows, bushes by a river, pond, or creek. This imaginary of the forest is key to what such a house is. This home is not part of the settlement of a people, and no other homes are ever to be found around it: instead, it destines itself to be part of the settlement of trees. The ecological mode in the imaginary of the home is especially pronounced in Efros's film, where the mother nurtures and heals wounded animals (that is, until her son arrives and hunts down a young deer). The forest extends the home into the whole of Earth, conceived as green freedom. As such, this space of outwardness also has a historical axis that places adjacency to the forest in relation to ecological formations of power.

Imagining forest as a space of freedom and of potency has clear roots in Russian history. Forest in the European part of Russia starts north of Tula, a town around three hundred kilometers south of Moscow. Muscovites in the Middle Ages were called the people "from beyond the forest." Below the forest, all the way south to the Black Sea and east to Siberia, lies an extremely fertile region known as black soil. The line dividing forest and steppe does not accurately align with the forest soil and black soil but nevertheless loosely corresponds to the political lines: first of medieval battles with the Golden Horde of nomadic people that populated the wide steppes of Eurasia in the thirteenth to fifteenth centuries and dwindled from then on; and secondly coinciding with the tideline of serfdom in Russia. In prehistoric and historic times, the forest stomped the steppe out, creeping south and seeding over cleared land, and the steppe encroached and devoured the forest by inviting its animals to graze there. In the fifteenth to seventeenth centuries, battling with its southern neighbors, the newly conceived state of Russia gave conquered black soil lands to landowners and the status of peasants was permanently fixed as that of serfs, partly to ensure that there was sufficient population to work those lands. The wonderful black soil requiring neither crop rotation nor other soil care became an exclusive site of slavery. This internal colonization, as Alexander Etkind has called it, is quite in line with external kinds.[55] The cotton-bearing fertile lands, the extreme productivity of sugar cane, and the exclusivity of growing certain spices are all part of the history of soil and plant-based colonialism.

In Russia it meant that the majority of the peasant population of the north (from Moscow to the White Sea) and of Siberia in the fifteenth to seventeenth centuries were either relatively free (state serfs) or had never been enslaved. Juridically, they were called "free countryside dwellers" and were personally free though often linked to their place of abode with a restriction placed on movement. The idea of the forest as coextensive with a free space that contains sources for survival but that does not demand or obligate grueling agrarian work, itself a condition of enslavement, lies here. What constructs this forest is relative scarcity, nonfertility. Part of the family of one of us comes from a village by a forest, near Archangel. This village consists of houses as if from Tarkovsky's films: by the river, and by the forest, not too near each other, nor too far apart, set in serene green and overpowered by midges. The father of one of us recounts that those peasants never became serfs simply because it was hard to survive, let alone sustain the wealth of another, making the "opportunity cost" of subjugation too substantial. Villagers relied on hunting, growing some vegetables, gathering berries and mushrooms, and fishing, with an occasional crop of rye. The forest was a site of survival, and little girls were taught to handle equipment and guns, and to orienteer in the woods alongside little boys. The forest's northern climate and hardship made possible some personal freedom and a certain looseness of fixed gender roles. The freedom to get lost in three or more pines, be eaten by wolves or starve, goes hand in hand with not being a serf, with not being heavily gendered. It is so hard to remain alive that one might as well be a bit free.

There is nothing essentially and a priori good about the forest. The forest was only a site of freedom because the grassy land, and especially the black soil, was the site of serfdom, and cultivation was an instrument of violence. Labor, and more specifically agrarian labor, was considered, in the work of Locke, a tool for the redemption of men.[56] It is by dropping sweat onto the soil that one makes it one's own. Using such Lockean justification of ownership, settler colonialism saw land that was not visibly farmed as appropriate for settlement. Here, practices of cultivation became a politics of planting. In the United States, Canada, and Australia, settler colonialism did not recognize indigenous forms of inhabitation and cultivation. Without visible European-style cultivation, which was taken to be civilizational progress, indigenous populations were deemed unenlightened and could therefore be dispossessed. Home is always related to land in the multiple ways in which land can be conceived. If in Russia working the land became the condition of serfdom, in colonies, not working the land in a *certain* manner became the condition of dispossession. In both cases, the home is negotiated in relation to attraction and repulsion—to cultivation and plants on land.

Some of the founders of modern political Zionism, influenced by European colonial thought, grounded the claim of the "people of the land of Palestine" through attachment to land, an agricultural colonization. In *The Conflict Shoreline,* Eyal Weizman and Fazal Sheikh document how the line of aridity in the Negev Desert functions as an instrument of dispossession.[57] Bedouins populating these areas are considered to be without legal ownership of land and are continuously uprooted. The legal ownership of the land is established on the basis of specifically formulated kinds of agricultural evidence: the intricate storing and usage of water and forms of cultivation of the desert by Bedouins are not recognized as such. A range of scholars have shown how—manipulating the "dead land" doctrine of the Ottoman land law, also upheld during British Mandate—the Israeli state transfers the Negev's "dead land" to the Israeli state for "revival."[58] Weizman writes that while the Bedouins cultivated along water streams, Israeli settlers intended to irrigate and cultivate the entire land surface, pushing the desert further south and in effect depleting water resources. Delivering water here is an act of political and ecological violence. The "natural force" of "making the desert bloom," argues Weizman, is part of Zionist imaginary. In relation to the Negev Desert, designating areas as "nature reserves" or "agricultural fields" is an act of colonial expropriation where ecological concerns are masks for active dispossession.[59] Weizman reports that for Yosef Weitz, the head of the Jewish National Fund's Lands and Forest Department, afforestation was imagined as "a biological declaration of Jewish sovereignty" and the generation of new geopolitical (and necessarily ecological) facts.[60] Spraying pesticides on Bedouin fields in acts of desertification on behalf of the Israeli state demonstrates further how the werewolf of the desert and the forest shifts and changes face but does not change the target of violence.[61]

The climate, argues Weizman, has always been a project for colonial powers, and the cultivation of plants an imperial force.[62] Plants were recruited into capitalism early: the engineer Joseph Paxton's glorious iron-structured glasshouses of the British gardens "back home" were the products of both the industrial age and imperial colonialism and housed jungle plants as a demonstration of wealth accrued overseas and as the sign of prestige—as well as a colonization of desire, where jungle represented endless growth and vitality. Today, planting a forest is not an a priori ecologically friendly gesture: it can be a violation or ruination. Brazilian farmers, wishing to extract more value from their land but unable to cut down protected rainforest, plant eucalyptus trees around it. Eucalyptus changes the acidity of the soil, which in turn kills rainforest flora.[63] The plant acts as a weapon and planting as extermination.

One being's forest, therefore, can be another being's desert. In the figuration of the transitory home in the forest, there is no cultivation: there is nowhere to cultivate, only to coexist with fir trees. The forest is not subjected to human rule through agrarian labor, a foundation of land and slave ownership. Rather, a forest is traversed, with its entire land surface remaining something of a mystery. While the European forest is still predominantly a temperate forest, the forest of Siberia, the taiga, is composed largely of conifers and extends all the way to the Pacific. This forest is home to extreme temperatures. One cannot plant an apple tree there with any hope. Firs stand for cold, scarcity, hunger—and, by extension, freedom. Perhaps to imagine a home as a space that is noncolonial, not-only-human—and that unites natal and cosmic, political and ecological, awayness and nurturing, while being outwardly open, occupying and giving space—one might start from a figuration of scarcity and freedom implied in such a northern forest. Others map such spaces in wilderness, in the highlands and marshlands of the world.[64] Those who tarry there may muse on whether they are just another trap.

ECONOMY–ECOLOGY

Many scholars, and Jacques Derrida in particular, have commented on the connection between ecology and economy to be found in the word *oikos* (ancient Greek for home, property, and family). This joint root does something to fuse them. In *oikos,* explains Cary Wolfe, relations between the self and the outside are normalized, distributed, and economically moderated.[65] *Oikos* unites the economic and ecological activities of the household. It is hard to disentangle shelter and value; it is tragic that it is hard to disentangle ecology and economy.

Peter Linebaugh's explication of the Charter of the Forest (a companion document to the 1215 Magna Carta), given in *The Magna Carta Manifesto,* highlights a British history of forests that is drastically different from those of what has become Russia. Here, between the twelfth and eighteenth centuries, forests—once held in common or commonly open to foraging and to the gathering of wood and plants providing subsistence to peasant populations—were enclosed.[66] Populations perished or migrated and the forests were destroyed with very few kept for reasons such as the entertainment to be found in hunting, as wood for the construction of ships, and as a scenic backdrop for large estates. Abolition of forms of common use and appropriation with further destruction of forests—jungle, wilderness, bush—is a familiar history in Europe and its colonies alike. The theft of commons and removal of personal liberty also went hand in hand globally.

At the threshold of "communal living upon natural resources held in common," for Linebaugh, lay various forms of enslavement.[67]

While the enclosure movement and slave trades destroyed forests and people in the making of industrial capitalism, Russia remained largely an agrarian country until the early twentieth century. Encyclopedias of that time mention that the area of state-owned Siberian forest is calculated only *approximately*. Its enslavement of its "own" citizens also had an economic foundation in labor, but while it is easy to assume that prior to capitalism or socialism, vast, unaccounted-for forests, open to everyone's use, retained some of the premodern ecological presence, economic independence, and freedom for humans, it would be romantically inaccurate. Humans engaged in wide-scale ecological destructions before and outside of the projects of agriculture and industrial capitalism.

Another look at Russian history can add different undertones to the idea of a free home in the forest. Russian children learn at school that fur skins, alongside cattle and early coins, were widely used as money or tax in Slavic settlements prior to the formation of states (prior to the ninth to tenth centuries). Etkind argues that the expansion of the Novgorod Republic (existing between the twelfth and fifteenth centuries) northward to the White Sea and eastward to the Ural Mountains, and prior to the expansion of the newly formed Muscovite state all the way eastward to Siberia and Alaska, was based on the fur trade. Etkind maintains that it is fur trade *above all* that financed the early mercantile democracy of Novgorod and the military campaigns of Ivan the Terrible alike,[68] becoming the biggest source of Russian state income from the tenth to the eighteenth century (depending on the historical source and the century, constituting between 10 and 25 percent), before collapsing completely in the nineteenth century. Russian fur was traded with Persia and Europe, where it was in high demand. Etkind reports that in one of Henry IV's outfits, London skinners reportedly used 12,000 squirrel and 80 ermine skins, and in the year 1391 alone, London imported 350,690 squirrel furs, up to 95 percent from Russia.[69]

In the sixteenth century, the new routes of trade with Northern Europe were established through the White Sea: with the depletion of the Russian forests and the decline of Novgorod, the Muscovite expansion to Siberia began, and in fur terms the emphasis shifted to sable, a species of marten.

Etkind follows a range of sources to make a claim that the expansion of the Russian Empire to Siberia was led entirely by fur and as late as the eighteenth century, a large part of the acquisition of the Hermitage Museum under the rule of Catherine II was financed by the state-controlled Siberian fur trade.[70] Sable furs were used as gifts in diplomatic relations (40 sables constituted a good gift) and there is evidence that officials of the Muscovite state sometimes received their salaries in fur.[71] In the seventy years between 1621 and 1690, Siberia delivered 7,248,000 sables.[72] Shortly after, their numbers went into great decline.

As Europe cut its woods, mythologizing the forest or poetically romanticizing the mountain top while appropriating pretty much the rest of the world's lands in its colonial effort, a range of commentators noted that the acquisition of land in the quest for fur by Russia was the largest ever accomplished due to one commodity.[73] In one such ecological history, developed in the nineteenth century by the historian Afanasiy Shchapov, "fishing colonization" is another legacy of expanding the state along waterways.[74] Etkind summarizes Shchapov's zoological economy as follows: "Beaver led the Russians to the place where they founded Novgorod; grey squirrel secured them the wealth of Moscow; sable led them to the place that became mapped as Siberia; sea otter brought them to Alaska and California."[75] This animal resource, exhausted in the eighteenth century and variably regulated from the beginning of the twentieth century until now, is another colonial story: where hunting, and not agrarian labor, is the foundation of wealth, the establishment of a claim of ownership and control over a territory, and a means of expansion.

Mobile, dynamic, not linked to one home but moving through lands and recruiting local populations into a globalizing fur trade, hunting practices are an undertone in the notion of home in the forest. Not on farmable land but in the forest full of wild animals, a house in the forest, a hunter's hut, can be wealthy, violent, and cruel. After the fur trade was discontinued, the areas of the north and east were used as ideal sites for penal colonies: so much for the site of freedom.

Home intertangles elements of human, animal, and plant life amid the nonorganic life of rocks. The destinies of species, politics, the length of winter, the type of soil, plants' identities, and humans' dreams are all stirred together but differentially, with varying powers spread among them. Gun, spade, snow, seed, humus, fur, fir, fungi, ammonia, bee. Can we think of home in the forest that is not indicative of colonization or privatization, not instrumental to hunting, to hunger or other extermination, not symbolic of imprisonment? The vector of animal hunting—land cultivation is an economic—ecological axis, a political and Marxist reading, of home that is always on Earth and has specific forms of materiality. With its spiritual existence, the material of home brings us back to the beginning of the chapter, to Bibikhin's forest, always capacious in its powers of abstraction and in concrete ability to intervene, to act, to prefigure, and to be hurt. The plants of serfdom are sociopolitically held at bay in this chapter's imagined home in the forest, and the animals are not made extinct by economic pursuits or due to raison-d'état. The metaphysical and real extrajuridical freedom of this forest is grounded in its various manifestations as matter, as the time of evolution, and as concrete ecologies. Its energy is no one's own property, and its concatenations are a place to live.

AT HOME IN THE FOREST OF MATTER

Ecological aesthetics has a passing relation to the uncanny in the way that the interrelation of things becomes a means of describing not holism but multiscalar connections that may at times introduce a strange frisson between things. But while the multiscalar nature of things can be regarded as simply factual (subatomic particles bearing numerous kinds of forces cohere as atoms, which manifest as molecules that aggregate to make substances that are in turn subject to other scales of forces, and so on, and that are potentially recursively interrogated and probed at every scale by ideas, instruments, and capacities that elicit the capacity to describe things as fact), things also have their limits. Connections may not always be welcome, and indeed may be hoped-for rather than achieved. In distressed states, which see cetaceans beaching themselves and cephalopods leaving the sea to traipse across the treacherous land to find a connection to calmer waters, the search for some kind of home is precarious. Things become uncanny when scales or conditions that are normally unrelated to a matter or to a problem are drawn in as a means of negotiating a situation and which thus cast them in a new light. The route through to a primal chance or to calmer waters is taken by setting off on a sometimes forlorn trek home: this is the experimental method in a tight corner. Being displaces itself in the indeterminacy of the forest. It is perhaps a means of finding an uncharted route through to another world, parallel to that won by means of abstraction, through scientific labor, through the forces of imagination and of practical action. How to navigate this world? What are the means of arriving at some coordinates?

Working in the middle of concrete ecological formations also means to work with politics as matter and poetics as a condition of possibility. Bibikhin talks about greater or lesser proximity to the raging bonfire of woods as being a determining characteristic of cultures. In the long now, in which the planet is being shaved of forest, the shaggy body of the forests are being given a fatal Brazilian; fuel is being extracted not from spatially but from temporally distant places, fossil forests in the form of oil and coal. The forests figure in Bibikhin as a philosophical "primary substance"—originating stuff—one that stands in for all matter and gives it its name. The forests are also ideas, myths, evolutionary forces, bacterial battles and mergers, cell adaptation and animal migration. They are also concrete forests, jungles, and the earth's life forms and the history of what has been done to them, and they include the perpetrators in their structurations.

Living in proximity to the forest, we propose, means living in a thoughtful and experimental relationship to matter and to movements of energy of life that are no longer solely primal but worked, elaborated, forgotten, ignored, distilled, alloyed, blocked, and layered with logics. Home in the forest of matter connects the configurations of the ethical and biological, political and poetic, ecological and social, physico-chemical, organic, historical, and other scales, brings them into conflict, and cuts through them. "Home on planet Earth" can only be imagined and made for living by recognizing the legacy of violence, both cruel and loving, by easing the grip of genus, by dethroning capital, by moving away from one's ordinary home, by creating and tuning into other sensibilities, actions, methods, *metis,* and forms.

To live at home in the forest, one may engage in material acts of experimentation that accrete as technologies—one may, indeed most likely will, live in the city, where the results of these have accumulated and where numerous flights from storms and from homelands have haphazardly or intentionally arrived or taken place. It is to become an ally with microbial consortia that are able to use plastic waste as a source of energy and carbon. Taking samples from a recycling site, a team of scientists in Japan recently found a strain of bacteria (Ideonella sakaiensis 201-F6) producing enzymes that metabolize polyethylene terephthalate, PET, the substance from which the majority of plastic bottles are produced.[76] Such work involves forming alliances between institutions, ideas, microscopes, databases, refuse workers, linguistic resources, and sludge. It requires finding means to trace the ways in which bacteria finds itself in the middle of things and elicits resources from the unpreposessing surfaces presented to it. Plastic-digesting bacteria may be a wonder for certain crises of refuse but a catastrophe for things such as surgical implants. Navigation and coordination in the midst of such things are a way of accreting a home that is away and outward and that necessarily comes into being across scales, articulating relations among the forest of matter.

Coda

Often, part of the function of a closing section of a book is to set out its findings in capsule form as an executioner's summary, a way of finishing it off. It may also be a word to the wise, one of caution in understanding and implementation, to ward off excessive passion in interpretation or use. The epilogue to the *Kama Sutra* is like this, where a note of moderation plays the role of a lull or a shift of register after a crescendo—the classic form of a coda. Such sage notes may read more like a disclaimer, shifting responsibility onto the reader without warranty, as do, at another pole, many books that end with an exhortation to revolutionary action, self-transformation, or the pursuit of the vista of the open road. Another approach is to hint at a possible unboxing of the underlying parcel of tricks, the conceptual gearing that underlies the mummery of the preceding chapters, or a glossary that puts it all right in the end, or at least gives the authors the chance to figure out what it is they have said by articulating it in a different register. It is at this point that a book may become stranger still, engaging in long circumambulations around a runaway idea that it has not yet quite gotten a fix on but that still provides its underlying core, one that is just hoving into view as it evaporates.

Every book is also something of a Trojan horse. Each carries its ostensive payload and outer form, as well as all of those things that travel along with it, even when they are not directly known by the authors. These payloads are not necessarily of any esoteric depth or cunningly devised mechanism of inscription ready to be decoded by the dutiful attention of whirring machines. They may be thought through theory while being born out of decades of bearing an ontological load, not fully coherent, static, or translatable into books. Such loads may include diary fragments in the form of their sense and source of their writing, as well as incidental interlopers and stowaways. They may be memories, dramas, experiences, visions with slippery fingers that try to grapple with ideological and biological burrs caught in the fur of their own shaggy dog stories. Experiences and formalisms cling to ideas as fragments of DNA might to the damply tangy surface of an overused banknote. This quality of the book as a gathering of fathoms of moments and of translations of durational distress, thoughts as commensalists, parasites, and symbionts that may be related or unknown to each other is something that allows things to meet that might not otherwise have come into composition and recognize their unspoken affinities. At times, they are what brings a book together, and in such conjunctures, something, even if only the guttering flame of a misconception, is always at stake.

The hosts may be hard-won insights, nailed-in sensibilities, open wounds, and half-formed opinions, a pile of oily ideational potsherds that, in combination with other fragments of text, produce a strangely complete curvaceous vessel. Part of the idea of this book has been to give such things a presiding impetus, to set them up to spill out into a chapter each, where they might not otherwise get a look. Each of the chapters concerns a background thought, something implicit and core in the tonus of contemporary forms of life but generally underworked. The challenge is to change, displace, or burrow into this condition. Find the flexures implied in the conjunction of parts and trace some further lines—hyperextensions, spring-loaded trajectories of actualization that recompose the axes and continuums that weave and clot together to render becoming.

Monte Testaccio is a hill amassed inside the city walls established for Rome under Marcus Aurelius. Still standing, it consists of the remains of olive oil amphorae. These rotund rough clay containers, mainly shipped to the city from Spain, were used only once. Over the course of the first two and a half centuries of the Common Era, many hundreds of millions of fragments and many whole amphorae were deposited here. Needless to say, this is something of a novel ecology: a dump of used vessels becomes an archive, layers corresponding to the age of the city, with the surface of pots also carrying various epigraphs, including graffiti, goods-control records, and dates to mark transit and freshness. Imagine a software system with each of those sherds scanned and sorted for possible combination with every other one, a computationally massive problem. Instead of organizing and disposing of them by simple resemblance to whole pots, they are being arranged by other means, whose principles are perhaps aligned with those of the analytic cubism of Pablo Picasso and Georges Braque, whose canvases share their sallow dirt and terracotta pallet. They jitter and wiggle, grind on each other's surfaces, create visions: another logic for and from their use transpires, struggles to find its form of composition. It is a dancing mountain. Such dancing mountains can become earthquakes. Its dance is overcast and pungent, morbid, sending voluminous clouds of dust up into the air as it heaves and clunks and spins. The parts that had been organized by sedimentary layers are now finding new affinities and resemblances to perform. The motley crew of cracks and gems, or fragments and beauties, lurches and settles, then lurches again.

This book has emphasized tendencies toward states of being, force fields, or conditions that do not yet imply particular kinds of subject as their ideal scale or that exist at multiple intermediary scales. These conditions in turn are not presented as simply natural phenomena but as things that emerge and shift over time, that have histories coursing through them. Aesthetic processes are understood as occurring dialogically, in a plurality of phrasings and forces, with multiple kinds of cut and temporality. For this reason, *Bleak Joys* brings together a set of asymmetrical entities including modes of individuation, sociotechnical grammars, and political propositions for expanded senses and conditions of aesthetics. Each of these attend to certain states that tend to be thrown into the shade by entities and processes that appear to be more significant from established loci of articulation.

One of these is part of the apparatus of the conceptual division of labor operative in the discipline of philosophy. In recent years there has been some welcome renewal of metaphysics as a means of finding new grounds for critical and speculative thinking. If we can, so the implication goes, refigure our understanding of the forces and conditions that compose the earth, of being and of the cosmos, perhaps new possibilities might come to light "downstream" for understanding ethics, aesthetics, politics, and so on. Given an attuned cosmic compass, we might even get to stumble out of this mess into another one. (There is always the possibility that the metaphysical entrails might be read as a way of avoiding such things, as a retreat from the world, but except for a few obvious cases this seems an unlikely gambit, given the necessity of such work.)

A metaphysics that might grow to provide a means of regrounding political sensibility and possibility in the present would, as one move, look at the conjunctions of categories of thought, assaying the way in which they rework each other. The compound form of the ethico-aesthetic developed via Bakhtin and Guattari is one result of this but one that also implies the possibility of a reattunement to metaphysical and thus to scientific concerns. The latter is a crucial aspect of such an approach, one that cannot be effaced by habitual compulsions to suspicion accrued as legacy in cultural theory. At the same time, there is something too presumptuous about the term "metaphysics," implying other sets of questions, other domains, as mere tributaries. Metaphysics can thus be read as a set of developing ways of addressing problems, dynamics, and entities, which has its own particular characteristics, and idiomatic means of interlacing with and inquiring into the operations and exigencies of others to which it may be more or less apt. Others may be more adequate to particular kinds of probe or question. The mode of individuation of a particular scale—say that of literature, a species, a gesture, or an event yet without a name—arises because it passes a threshold of becoming in which an irreducible consistency is reached. Each then requires idiomatic modes and instruments of understanding and inquiry, which tend to have reciprocal effects at other scales, generating abstractions or consequences that move between them. Abstract dynamics are thus generated by and move between both individuations, at multiple scales, and in ideas about them, as they are, in turn, formulated in different registers and idioms that may or may not respond to things, such as the capacities of different disciplines, currents, or capabilities of the imagination.

It is perhaps the difficulty of such a thing that has led some to wish to negate the Earth, to have done with it, at the same time as to establish an irrevocable binding to it. There are certain languages in which words such as "apocalypse" and "catastrophe" appear in every basic phrase book. They are perhaps learned too easily by publicists, who imagine themselves or their research outcomes to be the enormous meteorite that will finally and gratifyingly end it all. For others, the dissatisfaction is a little broader: there is a sense that all those other planets out there, although they are moderately unreachable, or rather inhospitable, must have one among them that is better configured for some version of an "us," even if it is only one that allows the fantasies of domination and disconnection that pertain in the present to maintain their trajectory without consequences. Which moon of which planet can be found to allow the mephitic air of British suburban life to maintain itself as if it were in a pure line of genesis from the 1950s or the third decade of the present century? Which capsule planet will allow full blossoming of the human potential of the generation of the North American 1960s? And which archipelago of planets still waits eagerly for the arrival of the first cosmonaut settlers carrying the red-and-black banners of interplanetary sorority? The Earth is too small for all its monocultures and they in turn are too miniscule to encompass it, too small in their scabbings together of meaning to do more than crystallize a more general incomprehension. Every intelligence inversely invents its own reciprocal form of stupidity, which may be of immense dimensions.

All these concrescences are relatively arbitrary, as is more or less any system for the organization of life, but therein lies their cruciality. What are the torsions worked on and in the arbitrary to bring us to this point? How is chance organized? What necessity can be heard in the racket of being? In responding to this question, Gottfried Willhelm Leibniz traced the best of all possible worlds as a signature of God. Others will look for different identifying marks on creation in order to work out its operating principles. What is it about the way anguish is undergone or devastations occur that release some spore-like clues about the condition of the world? Farce or comedy? Dark matter or some execrable process, pulling the strings? How, given the way a root tendril of a certain plant moves through the stones, under the paving, might we invent or undergo a means to think, even if only in a more tenderly observant way?

Animals are strange compared to flowering plants in that they often engage in seduction of their own—rather than another—species in order to reproduce. Anthropologically speaking, from the arts of seduction and persuasion, of inspiration—and of learning what might be passing through the other's mind—there perhaps arose some slightly more systematic forms of thought, such as, in the human, the political or philosophical. But it is conceivable that something is missing in all this advance. Perhaps, like flowers, there might be a need to learn to seduce other species or, better, to be entranced by them and thus set off a related chain of events in the ways in which sense is made of the world. Plants provide some basic ideational resources for such a development. But perhaps the question has, at another scale, more to do with the process of speciation and the way this in turn implies approaches by which the patternings and possibilities of seductions between species and of their intelligences might be elaborated, shift, or engender other dynamics. And there is a concatenation of this condition: the overlapping of species and scales is made concrete in the formation of niches and habitats and in the adaptions to them. In a manner related to the way species may depend upon particular scales and, as discussed above, engender ideational domains that may arise in relation to specific scales with their own epistemic demands, there is also a sense in which such movements of seduction may move across and between ideas. Humans are mutant creatures that are a fragment in time and of a scale at which we are a part, but they are also infested by brilliant abstractions.

Those ideas triaged in this book divulge multiple surfaces to each other that might lead to various kinds of construal and imagination. Passages open up to patterns of concrescence out of which something stabilizes as an idea, sometimes taking the form of something like a chapter but also of the reticule of assonance and dissonance, questioning and reworking, musing and misunderstanding, in which most of the work is already done by other texts, ideas, and the requirement to have some kind of relationship to reality, as complex and difficult as the latter might be in terms of its substance and the working of interpretation. Among this spread of entities are things that are generally sorted into the heaped categories of the wrong or the bad. It is not our intention to make the frequent move of carrying out an inversion of their value, to make them stand in for the right and the good. Rather, an intent of *Bleak Joys* is to attend to the means of engendering capacities for recognizing, while being amid it an ecological polyvocality of being. To do so means to work on ways of being difficult as well as on the difficulties of being.

Notes

INTRODUCTION

1. Catherine Malabou, *Ontology of the Accident: An Essay on Destructive Plasticity* (Cambridge, UK: Polity, 2012).
2. Aleksandr Pushkin, *Eugene Onegin: A Novel in Verse*, trans. Vladimir Nabokov, vol. 2, Bollingen Series 72 (Princeton, N.J.: Princeton University Press, 1981), 141. Nabokov's remarks on *toska* are in his commentary to chapter 1, stanza XXXIV, line 8.
3. Varlam Shalamov, *Gulag Tales*, trans. John Glad (London: Penguin, 1994).
4. Eugene Thacker, *After Life* (Chicago: University of Chicago Press, 2010).
5. Gregory Bateson, *Steps to an Ecology of Mind* (1972; repr., Chicago: University of Chicago Press, 2000).
6. Sarah Kane, "Cleansed," in *Complete Plays: Blasted; Phaedra's Love; Cleansed; Crave; 4.48 Psychosis; Skin* (London: Methuen, 2001), 136.
7. Jean Baudrillard, *Seduction* (London: St Martin's, 1991).
8. Anthony Trewavas, *Plant Behaviour and Intelligence* (Oxford: Oxford University Press, 2014); Frantisek Baluška, *Communication in Plants: Neuronal Aspects of Plant Life* (Heidelberg: Springer, 2016).
9. This is a commonplace in Soviet pop songs, Soviet science fiction, and, in some way, Russian culture more broadly.
10. V. V. Bibikhin, *Les (hyle): Problema materii, istorija poniatija, zhivaja materija v antichnoj i sovremennoj filosofii* [Forest (hyle): The problem of matter, history of concept, and living matter in ancient and contemporary biology] (St. Petersbourg: Nauka, 2011). Chapters 1, 6, and 8 are published in English as "The Wood(s) on the Problem of Living Matter in Ancient and Contemporary Biology," trans. Michael Marder, *Stasis*, no. 1 (July 4, 2015), http://www.stasisjournal.net.
11. Félix Guattari, *Chaosmosis: An Ethico-aesthetic Paradigm*, trans. Paul Bains and Julian Pefanis (Sydney: Power Publications, 1995).

12. For the development of a related argument, see Maurizio Lazzarato, "The Aesthetic Paradigm," in *Deleuze, Guattari and the Production of the New,* ed. Simon O'Sullivan and Stephen Zepke (London: Continuum, 2008), 173–83.

13. Félix Guattari, *Cartographies Schizoanalytiques* (Paris: Éditions Gallilée, 1989), 299; M. Bakhtin, *Voprosy literatury i estetiki,* ed. S. Leibovich (Moscow: Khudozhestvennaia literatura, 1975); French translation, Mikhail Bakhtine, *Esthétique et théorie du roman* (Paris: Gallimard, 1978); partial English translation, M. M. Bakhtin, *The Dialogic Imagination: Four Essays,* ed. Michael Holquist, trans. Caryl Emerson and Michael Holquist (Austin: University of Texas Press, 1981).

14. Guattari, *Chaosmosis,* 15. See also Mikhail Bakhtin, "Content, Material and Form in Verbal Art," in *Art and Answerability: Early Philosophical Essays,* ed. Michael Holquist and Vadim Liapunov, trans. Vadim Liapunov and Kenneth Brostrom (1924; repr., Austin: University of Texas Press, 1990), 257–325. The preceding points on the list are: (1) the sonority of the word, its musical aspect; (2) its material significations with their nuances and variants; (3) its verbal connections; (4) its emotional, intonational, and volitional aspects.

15. Bakhtin, *Art and Answerability,* 1.

16. Bakhtin, 14.

17. Maurizio Lazzarato, "Dialogism and Polyphony," in *No Ghost Just a Shell,* ed. Pierre Huyghe and Philippe Parrenno (Cologne: Walther König, 2003), 58–73.

DEVASTATION

1. Félix Guattari, *The Three Ecologies,* trans. Gary Genosko (London: Athlone, 2000). Guattari touches on questions related to devastation when he discusses pollution, algal blooms, and Donald Trump (43).

2. Gilles Deleuze, "La conception de la différence chez Bergson," *Etudes Bergsoniennes* 4 (1956): 77–112; translated as "Bergson's Conception of Difference," in *Desert Islands and Other Texts, 1953–1974,* ed. David Lapoujade, trans. Michael Taormina (Los Angeles: Semiotext(e), 2004), 32–51.

3. Alfred North Whitehead, *Modes of Thought* (New York: Free Press, 1968), 83.

4. Gilles Deleuze, *Difference and Repetition,* trans. Paul Patton (London: Athlone, 1994).

5. Benjamin Noys, *The Persistence of the Negative* (Edinburgh: University of Edinburgh Press, 2010).

6. Gilles Deleuze, *Pure Immanence: Essays on a Life* (New York: Zone Books, 2012).

7. Timothy A. Mousseau, Gennadi Milinevsky, Jane Kenney-Hunt, and Anders Pape Møller, "Highly Reduced Mass Loss Rates and Increased Litter Layer in Radioactively Contaminated Areas," *Oecologia* 175, no. 1 (May 2014): 429–37.

8. Simondon quoted in Jean-Hugues Barthélémy, "'Du mort qui saisit le vif': Simondonian Ontology Today," *Parrhesia: A Journal of Critical Philosophy* 7 (2009): 28.

9. Barthélémy, 32.

10. Ray Brassier, *Nihil Unbound: Enlightenment and Extinction* (London: Palgrave Macmillan, 2007), 237–38.

11. On endosymbiotic becoming, see Luciana Parisi, *Abstract Sex: Philosophy, Bio-Technology and the Mutations of Desire* (London: Continuum, 2004).

12. Donna Haraway, *When Species Meet* (Minneapolis: University of Minnesota Press, 2007), 256.

13. Lars von Trier, *Melancholia* (Copenhagen: Zentropa, 2011).

14. Eugene Thacker, *In the Dust of This Planet: The Horror of Philosophy*, vol. 1 (Winchester, UK: Zero Books, 2013).

15. The precautionary principle requires that the onus is on those making an intervention to ascertain that it will not negatively affect its ecological context before making it. See Tim O'Riordan and James Cameron, *Interpreting the Precautionary Principle* (London: Earthscan, 1994).

16. Artists too attempted to turn peoples' gaze back on the physical event. Notably, artist group Übermorgen declared the spill the largest oil painting in history, reframing aerial pictures of the event in such terms.

17. For an overview, see the website of Oilwatch, "a network of resistance to oil activities in tropical countries": http://www.oilwatch.org/.

18. Catherine Malabou, *Ontology of the Accident: An Essay on Destructive Plasticity*, trans. Carolyn Shread (Cambridge, UK: Polity, 2013), 73.

19. Malabou, 75, 81. Brassier with his project of nihilism reads Nietzsche and puts the operation of affirmation into doubt. He proposes a discussion of negativity via the transcendental scope of extinction (through solar death and extinction of abstraction). Brassier, *Nihil Unbound*.

20. Malabou, *Ontology of the Accident*, 89.

21. Elinor Ostrom, *Governing the Commons: The Evolution of Institutions for Collective Action* (Cambridge: Cambridge University Press, 1990).

22. Richard C. Thompson, Yiva Olsen, Richard P. Mitchell, Anthony Davis, Steven J. Rowland, Anthony W. John, Daniel McGonigle, and Andrea E. Russell, "Lost at Sea: Where Is All the Plastic?," *Science* 304, no. 5672 (2004): 838.

23. Andrés Cózar, Elisa Martí, Carlos M. Duarte, Juan García-de-Lomas, Erik van Sebille, Thomas J. Ballatore, Victor M. Eguíluz, J. Ignacio González-Gordillo, Maria L. Pedrotti, Fidel Echevarría, Romain Troublè, and Xabier Irigoien, "The Arctic Ocean as a Dead End for Floating Plastics in the North Atlantic Branch of the Thermohaline Circulation," *Science Advances* 3, no. 4 (April 5, 2017); Marcus Eriksen, Laurent C. M. Lebreton, Henry S. Carson, Martin Thiel, Charles J. Moore, José C. Borerro, François Galgani, Peter G. Ryan, and Julia Reisser, "Plastic Pollution in the World's Oceans: More than 5 Trillion Plastic Pieces Weighing over 250,000 Tons Afloat at Sea," *PLoS ONE* 9, no. 12 (2014).

24. Yukie Mato, Tomohiko Isobe, Hideshige Takada, Haruyuki Kanehiro, Chiyoko Ohtake, and Tsuguchika Kaminuma, "Plastic Resin Pellets as a Transport Medium for Toxic Chemicals in the Marine Environment," *Environmental Science and Technology* 35, no. 2 (2001): 318–24.

25. Here there is a significant differentiation from the gestation process of ambergris that forms a mass around the remains of undigested squid beaks in the bellies of sperm wales but that can in turn, it is proposed, possibly lead to the death of the whale. See Christopher Kemp, *Floating Gold: A Natural (and Unnatural) History of Ambergris* (Chicago: University of Chicago Press, 2012).

26. Amanda McCormick, Timothy J. Hoellein, Sherri A. Mason, Joseph Schluep, and John J. Kelly, "Microplastic Is an Abundant and Distinct Microbial Habitat in an Urban River," *Environmental Science and Technology* 48, no. 20 (2014): 11863–71.

27. Louiza Odysseos, "Human Rights, Self-Formation and Resistance in Struggles against Disposability: Grounding Foucault's 'Theorizing Practice' of Counter-Conduct in Bhopal," *Global Society* 30, no. 2 (2016): 179–200; Suroopa Mukherjee, *Surviving Bhopal: Dancing Bodies, Written Texts, and Oral Testimonials of Women in the Wake of an Industrial Disaster* (New York: Palgrave Macmillan, 2010).

28. YoHa [Harwood and Yokokoji], *Coal Fired Computers* (Newcastle-upon-Tyne: Discovery Museum, 2010). An accompanying booklet is available online: http://download.yoha.co.uk/cfc/CFC-200.pdf.

29. Boris Strugatsky and Arkady Strugatsky, *Roadside Picnic* (London: Gollancz, 2012).

30. Andrei Tarkovsky, dir., *Stalker* (Moscow: Mosfilm, 1979).

31. Malabou, *Ontology of the Accident*.

32. Paul N. Edwards, *A Vast Machine: Computer Models, Climate Data and the Politics of Global Warming* (Cambridge, Mass.: MIT Press, 2010); Jennifer Gabrys, *Program Earth: Environmental Sensing and the Making of a Computational Planet* (Minneapolis: University of Minnesota Press, 2016).

33. See Mary-Jane Rubenstein, *Worlds without End: The Many Lives of the Multiverse* (New York: Columbia University Press, 2014).

34. Gilles Deleuze, *Nietzsche and Philosophy* (London: Continuum, 2005); Bernard Stiegler, *Technics and Time 1: The Fault of Epimetheus* (Stanford: Stanford University Press, 1998).

35. Michel Serres, *Geometry*, trans. Randolph Burks (London: Bloomsbury, 2017).

36. Carl von Clausewitz, *On War*, trans. Michael Howard and Peter Paret, abridged by Beatrice Heuser (Oxford: Oxford University Press, 2007), 46. The question of the active witness is crucial here: "War is the realm of uncertainty; three quarters of the factors on which action in war is based are wrapped in a fog of greater or lesser uncertainty. A sensitive and discriminating judgement is called for; a skilled intelligence to scent out the truth" (46).

37. Alfred Jarry, *Exploits and Opinions of Dr. Faustroll Pataphysician*, trans. Simon Watson Taylor (Boston: Exact Change, 1996).

38. Appian, *Roman History*, book 8, part 1, *The Punic Wars*, Loeb Classical Library (Cambridge, Mass.: Harvard University Press, 1989).

39. Rachel Carson, *Silent Spring* (1962; repr., London: Penguin, 2000), 52.

40. Qiao Liang and Wang Xiangsui, *Unrestricted Warfare* (Beijing: PLA Literature and Arts Publishing House, 1999).

41. United Nations Convention to Combat Desertification, *Global Land Outlook* (Bohn: Secretariat of the United Nations Convention to Combat Desertification, 2017).

42. Eyal Weizman, *The Least of All Possible Evils: Humanitarian Violence from Arendt to Gaza* (London: Verso, 2011).

43. Qiao and Wang, *Unrestricted Warfare*.

44. There is an extensive debate around the accuracy of algorithms for the calculation of extinction rates for existing species and for the calculation of extinction rates for as yet undiscovered species. See, for instance, Fangliang He and Stephen P. Hubbell, "Species-Area Relationships Always Overestimate Extinction Rates from Habitat Loss," *Nature* 473, no. 7347 (May 19, 2011): 368–71. The question of normal rates of extinction compared to that of diversification and a comparison of methods of estimating them are discussed in Jurriaan M. De Vos, Lucas N. Joppa, John L. Gittleman, Patrick R. Stephens, and Stuart L. Pimm, "Estimating the Normal Background Rate of Species Extinction," *Conservation Biology* 29, no. 2 (2014): 452–62.

45. See Timothy Morton, *Ecology without Nature: Rethinking Environmental Aesthetics* (Cambridge, Mass.: Harvard University Press, 2009).

46. Sarah Catherine Walpole, David Prieto-Merino, Phil Edwards, John Cleland, Gretchen Stevens, and Ian Roberts, "The Weight of Nations: An Estimation of Adult Human Biomass," *BMC Public Health* 12, no. 439 (2012).

47. Erin E. Kershaw and Jeffrey S. Flier, "Adipose Tissue as an Endocrine Organ," *Journal of Clinical Endocrinology and Metabolism* 89, no. 6 (2004): 2548–56.

48. Félix Guattari, *Schizoanalytic Cartographies*, trans. Andrew Goffey (London: Bloomsbury, 2013), 21. See also Alexander A. Bachmanov, Danielle R. Reed, Michael G. Tordoff, R. Arlen Price, and Gary K. Beauchamp, "Nutrient Preference and Diet-Induced Adiposity in C57BL/6ByJ and 129P3/J Mice," *Physiology and Behavior* 72, no. 4 (2001): 603–13.

49. Lauren Berlant, *Cruel Optimism* (Durham, N.C.: Duke University Press, 2011).

50. For a survey of the unequal distribution of fast food outlets in Norfolk, see Eva R. Maguire, Thomas Burgoine, and Pable Monsivais, "Area Deprivation and the Food Environment over Time: A Repeated Cross-Sectional Study on Takeaway Outlet Density and Supermarket Presence in Norfolk, UK, 1990–2008," *Health Place*, no. 33 (May 2015): 142–47. On interspecies and class relations, see

Ellen K. Silbergeld, *Chickenizing Farms and Food: How Industrial Meat Production Endangers Workers, Animals, and Consumers* (Baltimore: Johns Hopkins University Press, 2016).

51. Jacques Perreti, "Why Our Food Is Making Us Fat," *The Guardian*, June 11, 2012. See also the film *King Corn*, directed by Aaron Woolf, written by Aaron Woolf, Ian Cheney, Curtis Ellis, and Jeffrey K. Miller (London: Mosaic Films, 2007).

52. Berlant, *Cruel Optimism*, 115.

53. There are a significant number of studies of this phenomena; for an early example, see Janneche Utne Skaare, Jon Morten Tuveng, and Hans Andreas Sande, "Organochlorine Pesticides and Polychlorinated Biphenyls in Maternal Adipose Tissue, Blood, Milk, and Cord Blood from Mothers and Their Infants Living in Norway," *Archives of Environmental Contamination and Toxicology* 17, no. 1 (January 1988): 55–63.

54. In *Marx and Nature: A Red and Green Approach* (London: Palgrave Macmillan, 1999), Paul Burkett describes the problem in relation to the question of thermodynamics, where it is true that "if we have enough energy, we could even separate the cold molecules of a glass of water and assemble them into ice cubes," but "in practice . . . such operations are impossible . . . because they would require a practically infinite time" (62). This problem applies in particular to those "elements which, because of their nature and the mode in which they participate in the natural and man-conducted processes, are highly dissipative" and/or "found in very small supply in the environment" (63). In short, "the somber message of the second law (that dissipation of matter and energy are unavoidable consequences of their use) mutes the seemingly optimistic message of the first law (that matter and energy are not literally consumed in their use)" (64).

ANGUISH

1. Kira Muratova, dir., *Three Stories* (Russia and Ukraine: NTV-Profit, 1997).

2. Mark Fisher, *Ghosts of My Life: Writings on Depression, Hauntology and Lost Futures* (Winchester, UK: Zero Books, 2014).

3. Julia Kristeva, *Black Sun: Depression and Melancholia* (New York: Columbia University Press, 1989); Susan Sontag, *Under the Sign of Saturn* (New York: Penguin Books, 1980).

4. See Franco "Bifo" Berardi, *Félix Guattari: Thought, Friendship and Visionary Cartography* (London: Palgrave Macmillan, 2008).

5. Sianne Ngai, *Ugly Feelings* (Cambridge, Mass.: Harvard University Press, 2005).

6. Lauren Berlant, *Cruel Optimism* (Durham, N.C.: Duke University Press, 2011).

7. Gregory Bateson, *Steps to an Ecology of Mind: Collected Essays in Anthropology, Psychiatry, Evolution, and Epistemology* (Chicago: University of Chicago Press, 2000).

8. Aleksandr Pushkin, *Eugene Onegin: A Novel in Verse*, vol. 2, *Commentary and Index*, trans. Vladimir Nabokov (Princeton, N.J.: Princeton University Press, 1990), xxxiv.

9. Malabou explicitly writes about the difficulties of creating an ontology of the accident, as it is of a character of a law but one that cannot be anticipated. If an accident occurs, it does not respond to its own necessity and does not "come to pass." Catherine Malabou, *Ontology of the Accident: An Essay on Destructive Plasticity*, trans. Carolyn Shread (Cambridge, UK: Polity, 2013), 30.

10. Malabou, 20–37.

11. Albert North Whitehead, *Process and Reality: An Essay in Cosmology* (New York: Free Press, 1985), 166, 167.

12. Steven Shaviro, *Without Criteria: Kant, Whitehead, Deleuze, and Aesthetics* (Cambridge, Mass.: MIT Press, 2012), 3.

13. Gilles Deleuze and Félix Guattari, *A Thousand Plateaus: Capitalism and Schizophrenia* (New York: Continuum, 2004), 11.

14. Gilles Deleuze, *Pure Immanence: Essays on a Life* (New York: Zone Books, 2001).

15. Clément Rosset, *Logique de pire: Éléments pour un philosophie tragique* (Paris: Presses Universitaire de France, 1979).

16. Gilles Deleuze, *Nietzsche and Philosophy* (New York: Continuum, 2005), 3.

17. Deleuze, 184–85.

18. Varlam Shalamov, *Kolyma Tales*, trans. John Glad (London: Penguin, 1994), 72.

19. Shalamov, "Dry Rations," in *Kolyma Tales*, translation ours from handwritten manuscript published online, https://shalamov.ru/manuscripts/text/15/38.html.

20. Shalamov, *Kolyma Tales*, 146.

21. Friedrich Nietzsche, *The Birth of Tragedy and Other Writings*, trans. Ronald Speirs (Cambridge: Cambridge University Press, 2006).

22. Friedrich Nietzsche, *The Gay Science*, trans. Thomas Common (New York: Dover, 2006), 46–47.

23. Nietzsche, *The Gay Science*, 22, 31.

24. Nietzsche, 47.

25. Deleuze, *Nietzsche and Philosophy*, 121.

26. Deleuze, 16.

27. Deleuze, 21.

28. Deleuze, *Pure Immanence*, 84.

29. Deleuze, *Nietzsche and Philosophy*, 7.

30. Rob Nixon, *Slow Violence and the Environmentalism of the Poor* (Cambridge, Mass.: Harvard University Press, 2013).

31. Scott Forbes, *A Natural History of Families* (Princeton, N.J.: Princeton University Press, 2005).

32. Forbes, 154. See also Murdock K. McAllister and Bernard D. Roitberg, "Adaptive Suicidal Behaviour in Pea Aphids," *Nature* 328, no. 6133 (August 27, 1987): 797–99.

33. Mikhail Gasparov, *Zapisi i vypiski* (Moscow: NLO, 2001), 82, translation ours.

34. Dmitry Bykov, "Interview with Kira Muratova," 1997, http://ru-bykov.livejournal.com/980096.html.

35. Christa Wolf, *Parting from Phantoms: Selected Writings, 1990–1994*, ed. and trans. Jan van Heurck (Chicago: University of Chicago Press, 1998), 199.

36. Deleuze, *Pure Immanence*, 26–28.

37. Gilles Deleuze, *Expressionism in Philosophy: Spinoza* (New York: Zone Books, 1992), 173–74.

38. Eugene Thacker, *After Life* (Chicago: University of Chicago Press, 2010), 27, 137, 144, 209, 212–34.

39. Thacker, 213.

40. Thacker, 226.

41. Thacker, 227.

42. Thacker, 151, 213, 214.

43. Thacker, 228.

44. Thacker, 233.

45. Eva Dien Brine Markvoort, *65 Red Roses*, accessed April 1, 2019, https://65redroses.livejournal.com/.

46. This saying is repeatedly cited and attributed to Sergei Averinstev, who is said to often use this sentence in his lectures.

IRRESOLVABILITY

1. David Markson, *The Last Novel* (Berkeley: Counterpoint, 2007), 63.

2. Arthur Schopenhauer, *The Essays of Arthur Schopenhauer: The Art of Literature*, trans. T. Bailey Saunders (London, 1891; Project Gutenberg, 2004), http://www.gutenberg.org.

3. Christa Wolf, *City of Angels* (New York: Farrar, Straus and Giroux, 2013), 289–90. See also Christa Wolf, *Cassandra: A Novel and Four Essays* (London: Virago, 1984).

4. G. W. F. Hegel, *Phenomenology of Spirit*, trans. A. V. Miller (Oxford: Oxford University Press, 1977), 7.

5. John von Neumann, *Theory of Self-Reproducing Automata* (Urbana: University of Illinois Press, 1967).

6. Intergovernmental Panel on Climate Change, "AR5 Synthesis Report: Climate Change 2014," https://www.ipcc.ch.

7. Raymond Williams with Michael Orrom, *Preface to Film* (London: Film Drama, 1954); Raymond Williams, *The Long Revolution* (1961; repr., Cardigan: Parthian Books, 2011); Raymond Williams, *Marxism and Literature* (Oxford: Oxford University Press, 1977).

8. Mike Davis, *Ecology of Fear: Los Angeles and the Imagination of Disaster* (New York: Vintage Books, 1999).

9. Martin Heidegger, *Question Concerning Technology and Other Essays* (1949; repr., London: Harper Perennial, 1982).

10. Martin Heidegger, "Letter on Humanism," in *Basic Writings*, by Martin Heidegger, ed. David Farrell Krell (London: Routledge, 2004), 223.

11. Oskar Morgenstern and John von Neumann, *Theory of Games and Economic Behaviour* (Princeton, N.J.: Princeton University Press, 1944).

12. Philip Mirowski, *Machine Dream: Economics Becomes a Cyborg Science* (Cambridge: Cambridge University Press, 2002).

13. Raymond Williams, "The Welsh Industrial Novel," in *Culture and Materialism: Selected Essays* (1980; repr., London: Verso, 2005), 213–31.

14. John Nash, "Non-Cooperative Games" (PhD diss., Princeton University, 1950); John Nash, "Equilibrium Points in N-Person Games," *Proceedings of the National Academy of Sciences* 36, no. 1 (1950): 48–49; John Nash, "Non-Cooperative Games," *Annals of Mathematics* 54, no. 2 (1951): 286–95.

15. Thomas Schelling, *The Strategy of Conflict*, 2nd ed. (Cambridge, Mass.: Harvard University Press, 1980).

16. Brian Massumi, "National Enterprise Emergency: Steps towards an Ecology of Powers," *Theory, Culture and Society* 26, no. 6 (2009): 153–85.

17. Bertrand Russell, *Common Sense and Nuclear Warfare* (1959; repr., London: Routledge, 2010).

18. Jean-Paul Sartre, *Existentialism and Humanism*, trans. Philip Mairet (1946; repr., London: Methuen, 2007). Perhaps, in the then emerging context of the Cold War, Sartre's indefatigable loyalty to the USSR—alternately personified as Stalin or of the image of the movement of global proletariat—can be seen in the light of the question of choice, to be like Kierkegaard's Abraham, in the glory of being a "father of faith" (a term also used in "Speech in Praise of Abraham") or indeed as the inventor of the notion of faith, having stuck by an impossible choice. While Sartre's idea of decision may take on the enlightening garb of the Hegelian sunburst of reason, the virtue of his position is that he denaturalizes the human and, as a consequence, also denaturalizes history and its passage. This corresponds, for instance, to the grounds on which Sartre criticized Albert Camus's *Le Peste*, for comparing Nazism to a mere plague. Søren Kierkegaard, *Fear and Trembling*, trans. Alastair Hannay (London: Penguin, 1986); Albert Camus, *The Plague*, trans. Robin Buss (London: Penguin, 2002).

19. See Jean-Paul Sartre, *The Family Idiot: Gustave Flaubert 1821–1857*, vol. 3, part 2, "Personalization," trans. Carol Cosman (1971; repr., Chicago: University of Chicago Press, 1989).

20. Gregory Bateson, *Steps to an Ecology of Mind: Collected Essays in Anthropology, Psychiatry, Evolution, and Epistemology* (Chicago: University of Chicago Press, 2000).

21. Lisa Blackman, "Affect and Automaticity: Towards an Analytics of Experimentation," *Subjectivity*, no. 7 (2014): 362–84.

22. Blackman discusses Benjamin Libet's formulation of a half-second delay between nervous activation and conscious volition in this light. Benjamin, Libet, Curtis A. Gleason, Elwood W. Wright, and Dennis K. Pearl, "Time of Conscious Intention to Act in Relation to Onset of Cerebral Activity (Readiness-Potential): The Unconscious Initiation of a Freely Voluntary Act," *Brain* 106, no. 3 (1983): 623–42.

23. For footage of Boshier talking about this painting, see Ken Russell's 1962 documentary for the BBC's Monitor series, *Pop Goes the Easel*. Echoing its geopolitical dimension, the painting has been owned since 1975 by Muzeum Sztuki in Łódź.

24. The toothpaste image features, like the abundance and whiteness of milk, in some of Boshier's other paintings.

25. Kurt K. Goedel, "On Undecidable Propositions of Formal Mathematical Systems" (notes by S. C. Kleene and J. B. Rosser on lectures at the Institute for Advanced Study, Princeton, N.J., 1934), in *Kurt Goedel: Collected Works*, vol. 1, ed. Solomon Feferman (Oxford: Oxford University Press, 1986), 346–71; Philip Mirowski, *Machine Dreams: Economics Becomes a Cyborg Science* (Cambridge: Cambridge University Press, 2002), 118.

26. Mirowski, *Machine Dreams*, 548.

27. Gilles Deleuze, *Foucault* (London: Athlone, 1988), 7, 18, 55; Matthew Fuller and Graham Harwood, "Abstract Urbanism," in *Code and the City*, ed. Rob Kitchin and Sung-Yueh Perng (London: Routledge, 2016), 61–71.

28. Deleuze, *Foucault*, 9.

29. Karl Marx, *Capital: Volume 1*, trans. Ben Fowkes (London: Penguin, 1990), 899.

30. Harry Magdoff and Paul M. Sweezy, *Stagnation and the Financial Explosion: Economic History as It Happened* (New York: Monthly Review Press, 1987).

31. Louise Amoore, *The Politics of Possibility: Risk and Security beyond Probability* (Durham, N.C.: Duke University Press, 2013), 46.

32. Brian Massumi, "Potential Politics and the Primacy of Preemption," *Theory and Event* 10, no. 2 (2007). See also Brian Massumi, "National Enterprise Emergency: Steps Towards an Ecology of Powers," *Theory, Culture and Society* 26, no. 6 (2009): 153–85.

33. Eric Schlosser, *Command and Control* (London: Penguin, 2014).

34. Arjun Appadurai, *The Fear of Small Numbers: An Essay on the Geography of Anger* (Durham, N.C.: Duke University Press, 2006); Bertrand Russell, *Common Sense and Nuclear Warfare* (London: Routledge, 2009), 31–32.

35. See Susan Schuppli, *Material Witness* (Cambridge, Mass.: MIT Press, forthcoming), for a virtuoso discussion of the magnetic tape.

36. Claude E. Shannon, "Von Neumann's Contribution to Automata Theory," *Bulletin of the American Mathematical Society* 64 (1958): 123–29. See also John von Neumann, "Probablistic Logic and the Synthesis of Reliable Organisms from Unreliable Components," in *Automata Studies*, ed. Claude Shannon and John McCarthy (Princeton, N.J.: Princeton University Press, 1956), 43–98.

37. Friedrich Hayek, "Economics and Knowledge," in *Individualism and Economic Order* (Chicago: University of Chicago Press, 1948). One of the aspects of Hayek's work that is attractive is that he reckons with qualities of highly differentiated kinds of knowledge, intelligence, and reason, which might in other vocabularies be called "situated," rather than positing a single form of rational self-interested individual. Reason, however, is determined by the interaction with the market, which introduces a uniform kind of rationality. Simon Griffiths makes an interesting reading of Hilary Wainwright's argument for such knowledge also being socially generated and shared, leaving it open to cooperative forms. See Simon Griffiths, *Engaging Enemies: Hayek and the Left* (London: Rowman and Littlefield International, 2014); Hilary Wainwright, *Arguments for a New Left: Answering the Free-Market Right* (Oxford: Blackwell, 1994).

38. F. A. Hayek, *The Sensory Order: An Inquiry into the Foundations of Theoretical Psychology* (Chicago: University of Chicago Press, 1952).

39. Heinz von Foerster and George W. Zopf, eds., *Principles of Self-Organization: Transactions of the University of Illinois Symposium on Self-Organization*, International Tracts in Computer Science and Technology and Their Application 9 (Oxford: Pergamon, 1962).

40. F. A. Hayek, *Law, Legislation and Liberty*, vol. 2, *The Mirage of Social Justice* (Chicago: University of Chicago Press, 1978).

41. Ludwig van Mises, *Human Action* (New Haven, Conn.: Yale University Press, 1949). In writings on catallaxy, economic theory becomes "catallactics," itself a term derived from Richard Whateley's *Introductory Lectures on Political Economy*, 2nd ed. (London: B. Fellowes, 1832).

42. John Ranelagh, *Thatcher's People: An Insider's Account of the Politics, the Power, and the Personalities* (London: HarperCollins, 1991); Jonathan D. Ostry, Prakash Loungani, and Davide Furceri, "Neoliberalism: Oversold?," *Finance and Development* 53, no. 2 (June 2016).

43. Friedrich von Hayek, *Law, Legislation and Liberty*, vol. 3, *The Political Order of a Free People* (Chicago: University of Chicago Press, 1981).

44. Hayek, 68.

45. Friedrich von Hayek, *The Road to Serfdom* (London: Routledge, 2001).

46. William E. Connolly, *The Fragility of Things: Self-Organizing Processes, Neoliberal Fantasies, and Democratic Activism* (Durham, N.C.: Duke University Press, 2013).

47. See, for instance, Thomas Piketty, *Capital in the Twenty-First Century* (Cambridge, Mass.: Harvard University Press, 2014).

48. Friedrich von Hayek, "Degrees of Explanation," *British Journal for the Philosophy of Science* 6, no. 23 (November 1955): 209–25; Friedrich von Hayek, "The Theory of Complex Phenomena," in *The Critical Approach to Science and Philosophy: Essays in Honor of Karl R. Popper*, ed. Mario A. Bunge (New York: Free Press of Glencoe, 1964).

49. Hayek, *Law, Legislation and Liberty*, 3:76.

50. Pierre Dardot and Christian Laval, *The Way of the World: On Neoliberal Society*, trans. Gregory Elliott (London: Verso, 2013).

51. David Lea, "'The Conscience of the World': Lars Iyer on Wittgenstein Jr.," *London Review Bookshop*, August 2014, https://www.londonreviewbookshop.co.uk. See also Lars Iyer, *Wittgenstein Jr.* (London: Melville House, 2014).

52. This is not to say that something like irresolvability does not arise in other conditions. The texture, speed, and impetus of difficulties during different early phases of the Soviet Union or in those parts of Spain run through anarchist and anarcho-syndicalist systems of workers' councils in the 1930s, for instance, would have been different in numerous ways but would also entail lacunae. One can see the decision-making processes of recent movements based around assemblies as engaging a form of automata (something taken up in the group decision-making software Loomio) that in turn makes certain kinds of problems more tractable. Gaston Leval, *Collectives in the Spanish Revolution*, trans. Vernon Richards (London: Freedom, 1975).

53. Grégoire Chamayou, *Drone Theory*, trans. Janet Lloyd (London: Penguin, 2014).

54. Nanni Balestrini, *Tristano: A Novel*, trans. Mike Harakis (London: Verso, 2014).

55. Umberto Eco, foreword to *Tristano: A Novel*, by Nanni Balestrini, trans. Mike Harakis (London: Verso, 2014), ix.

56. Sarah Kane, "Cleansed," in *Complete Plays: Blasted; Phaedra's Love; Cleansed; Crave; 4.48 Psychosis; Skin* (London: Methuen, 2001), 136.

57. Christa Wolf, *Medea: A Novel*, trans. John Cullen (London: Virago, 1998).

58. For instance, see Euripides, "Medea," in *Medea and Other Plays*, trans. Philip Vellacott (London: Penguin, 1963).

LUCK

1. Christopher N. Connolly, "Nerve Agents in Honey," *Science* 358, no. 6359 (October 2017): 38–39.

2. On flood prevention strategies, see David Harvey, *Spaces of Global Capitalism: Towards a Theory of Uneven Geographical Development* (London: Verso, 2006).

3. An entire legacy of modern and contemporary cultural currents could be set out on the basis of their relations to chance, the way chance is worked with or understood as a macro- or microcosmic substance in artworks or in dispositions

to life more generally. For one such discussion in relation to surrealism, taken as a cultural, philosophical, and political movement rather than one simply pertaining to art and literature, see Raihan Kadri with Michael Richardson and Krzystof Fijalkowski, "Objective Chance," in *Surrealism, Key Concepts*, ed. Michael Richardson and Krzystof Fijalkowski (London: Routledge, 2016), 143–53.

4. This is something set out well in the first section of Ulrich Beck's *Risk Society: Towards a New Modernity*, trans. Mark Ritter (London: SAGE, 1992).

5. Johann Huizinga, *Homo Ludens: A Study of the Play Element in Culture* (1939; repr., Boston: Beacon, 1955); Katie Salen and Eric Zimmerman, *The Rules of Play* (Cambridge, Mass.: MIT Press, 2003).

6. Gilles Deleuze, *Logic of Sense* (London: Continuum, 2004), 58; Lewis Carroll, "Alice's Adventures in Wonderland," in *The Annotated Alice: Alice's Adventures in Wonderland and Through the Looking-Glass*, by Lewis Carroll, ed. Martin Gardiner (London: Penguin, 2001).

7. A useful critical assessment of the apparent roots of these concepts in Stoic thought is made by John Sellars, "Aion and Chronos: Deleuze and the Stoic Theory of Time," *Collapse*, no. 3 (2007): 177–205. See also John Sellars, *Stoicism* (Durham, N.C.: Acumen, 2006). For a discussion of Chronos and Aion, see Richard Pinhas and Gilles Deleuze, "Cours Vincennes: On Music," May 3, 1977, trans. Timothy S. Murphy, https://www.webdeleuze.com.

8. Friedrich Nietzsche, *Thus Spoke Zarathustra*, trans. R. J. Hollingdale (London: Penguin, 2003), chap. 3, "The Seven Seals."

9. Deleuze, *Logic of Sense*, 63. Aion appears here as another form of Osiris-Dionysus.

10. Gilles Deleuze, *Difference and Repetition* (London: Athlone, 1994), 199.

11. Stéphane Mallarmé, *Un coup de des* (1897; repr., Paris: Nouvelle Revue Francais, 1914). See also Stéphane Mallarmé, *Collected Poems*, trans. Henry Weinfield (Berkeley: University of California Press, 1996).

12. On Zarathustra, see Gilles Deleuze, *Nietzsche and Philosophy* (London: Continuum, 1986), 28.

13. Franco Berardi, *Félix* (London: Palgrave Macmillan, 2008), 53.

14. Gilles Deleuze, *Coldness and Cruelty*, trans. Jean McNeil (New York: Zone Books, 1991).

15. Jean Baudrillard, *Seduction* (New York: St. Martin's, 1991), 133.

16. Terence Sellers, *The Correct Sadist* (Brighton: Temple, 1990).

17. Baudrillard, *Seduction*, 143.

18. Deleuze, *Nietzsche and Philosophy*, 26.

19. This is the kind of mad affirmation found in Bataille's introduction to his book on Nietzsche, part of the *somme atheologique*, of the war years and written in the frenzied closing months of 1944. See Georges Bataille, *On Nietzsche* (London: Athlone, 1992).

20. Clément Rosset, *Logique de pire: Éléments pour un philosophie tragique* (Paris: Presses Universitaire de France, 1979).

21. Deleuze, *Nietzsche and Philosophy*, 26–27.

22. Much of Naseem Nicholas Taleb's *Black Swan: The Impact of the Highly Improbable* (London: Penguin, 2007) is concerned with such matters.

23. V. V. Kozlov and M. Yu. Mitrofanova, "Galton Board," *Regular Chaotic Dynamics*, no. 8 (2002): 431–39. See also Asger Jorn, *Open Creation and Its Enemies*, trans. Fabian Tompsett (London: Unpopular Books, 1994).

24. On modeling, see Stuart N. Lane, Catharina Landstrom, and Sarah J. Whatmore, "Imagining Flood Futures: Risk Assessment and Management in Practice," *Philosophical Transactions of the Royal Society A* 369, no. 1942 (May 2011): 1784–806. On risk management, see Michael Power, *Organized Uncertainty: Designing a World of Risk Management* (Oxford: Oxford University Press, 2007). On the biopolitical force of statistics, see Ian Hacking, *The Taming of Chance* (Cambridge: Cambridge University Press, 1990); Michel Foucault, *The Birth of Biopolitics: Lectures at the Collège de France 1978–1979*, ed. Michel Senellart, trans. Graham Burchell (London: Palgrave Macmillan, 2007).

25. For a contribution to an account of such forms, see Matthew Fuller and Andrew Goffey, *Evil Media* (Cambridge, Mass.: MIT Press, 2012).

26. Andrey Platonov, "Fourteen Little Red Huts," in *The Portable Platonov* (Moscow: Glas, 1999), 158.

27. See Alexander Pushkin, "The Lay of the Wise Oleg," in *Anthology of Russian Literature from the Earliest Period to the Present Time*, vol. 2, *The Nineteenth Century*, ed. Leo Wiener (New York: Benjamin Blom, 1967), 139–42.

28. Robert K. Merton, *Social Theory and Social Structure* (New York: Free Press, 1968).

29. Alfred Cowles and Herbert E. Jones, "Some a Posteriori Probabilities in Stock Market Action," *Econometrica* 5 (1937): 280–94. See also Donald MacKenzie, *An Engine, Not a Camera: How Financial Models Shape Markets* (Cambridge, Mass.: MIT Press, 2006), 95.

30. Cowles and Jones, "Some a Posteriori Probabilities," 294.

31. William Burroughs, *Naked Lunch* (1959; repr., London: Paladin, 1986), 107.

32. Franco "Bifo" Berardi, *The Soul at Work: From Alienation to Autonomy*, trans. Francesca Cadel and Giuseppina Mechia (Los Angeles: Semiotext(e), 2009).

33. Susan George, *A Fate Worse than Debt* (London: Pelican, 1988).

34. Emily Martin, "The Egg and the Sperm: How Science Has Constructed a Romance Based on Stereotypical Male-Female Roles," *Signs* 16, no. 3 (Spring 1991): 485–501.

35. On Athenian democracy, see C. L. R James, "Every Cook Can Govern: A Study of Democracy in Ancient Greece; Its Meaning for Today," *Correspondence* 2, no. 12 (June 1956).

36. Arjun Appadurai, *Fear of Small Numbers* (Durham, N.C.: Duke University Press, 2006), 9.

37. For a discussion of the relation between computation and sorcery, see Florian Cramer, *Words Made Flesh* (Rotterdam: Piet Zwart Institute, 2005).

38. Scott Forbes, *A Natural History of Families* (Princeton, N.J.: Princeton University Press, 2007).

PLANT

1. Michael Marder, *Plant Thinking: A Philosophy of Vegetal Life* (New York: Columbia University Press, 2013); Jeffery Nealon, *Plant Theory: Biopower and Vegetable Life* (Stanford: Stanford University Press, 2015); Matthew Hall, *Plants as Persons: A Philosophical Botany* (New York: State University of New York Press, 2011); Anna Tsing, *The Mushroom at the End of the World: On the Possibility of Life in Postcapitalist Ruins* (Princeton, N.J.: Princeton University Press, 2015).

2. Phillipe Descola, *The Ecology of Others*, trans. Geneviève Godbout and Benjamin P. Luley (Chicago: Prickly Paradigm, 2013).

3. On gardens, see Robert Pogue Harrison, *Gardens: An Essay on the Human Condition* (Chicago: University of Chicago Press, 2008). On landscape, see Jason Orton and Ken Worpole, *The New English Landscape* (London: Field Station, 2013). On vegetal life, see Tsing, *The Mushroom at the End of the World*; Elaine Gan, *Rice Child (Stirrings)*, art installation, 2011–14, http://elainegan.com; Maria Thereza Alves, *Seeds of Change: A Floating Ballast Seeds Garden for Bristol*, art project, 2012–16, http://www.mariatherezaalves.org.

4. Gregory Bateson, *Mind and Nature: A Necessary Unity* (New York: Bantam, 1980). For a discussion of Beer, see Andrew Pickering, *The Cybernetic Brain: Sketches of Another Future* (Chicago: University of Chicago Press, 2011); Eduardo Kohn, *How Forests Think: Toward an Anthropology beyond the Human* (Berkeley: University of California Press, 2013). Beer's suggestion, in *A Platform for Change*, that a country pond or any other similarly complex self-regulating system would be able to substitute for the management layer of an organization is worth excavating. Stafford Beer, *A Platform for Change* (Hoboken, N.J.: Wiley, 1975).

5. Anthony Trewavas has been eminent in this discussion. See Anthony Trewavas, *Plant Behaviour and Intelligence* (Oxford: Oxford University Press, 2014).

6. Indeed, a move toward ICT-ethics is now imagined to supplant the boom in applied ethics, associated with genetic technologies. It is certain that we will see more of this, with philosophical thought and reflections on computing more broadly becoming a branch of actuarial calculation, with the simple difference that it is carried out in natural languages rather than formulae, though those will come too.

7. Isabelle Stengers's notes on the relation between neo-Darwinism and vitalism are of relevance here. See Isabelle Stengers, *Cosmopolitics*, 7 books in 2 vols. (Minneapolis: University of Minnesota Press, 2011), book 7, *The Curse of Tolerance*.

8. For a survey, see Fatima Cvrčková, Helena Lipavská, and Viktor Žárský, "Plant Intelligence: Why, Why Not or Where?," *Plant Signaling and Behavior* 4, no. 5 (May 2009): 394–99.

9. Trewavas, *Plant Behaviour and Intelligence*.

10. Trewavas (195) cites a catalog of seventy definitions of intelligence: Shane Legg and Marcus Hutter, "A Collection of Definitions of Intelligence," *Frontiers in Artificial Intelligence and Applications*, no. 157 (2007): 17–24.

11. Richard Karban, *Plant Sensing and Communication* (Chicago: University of Chicago Press, 2015).

12. Charles Darwin and Francis Darwin, *The Power of Movement in Plants* (London, 1880; Project Gutenberg, 2004), http://www.gutenberg.org.

13. Trewavas, *Plant Behaviour and Intelligence*, 132–33.

14. Karban, *Plant Sensing and Communication*.

15. František Baluška, Dieter Volkmann, Andrej Hlavacka, Stefano Mancuso, and Peter W. Barlow, "Neurobiological View of Plants and Their Body Plan," in *Communication in Plants*, ed. František Baluška, Stefano Mancuso, and Dieter Volkmann (Berlin: Springer, 2006), 19–35.

16. Gilles Deleuze and Félix Guattari, *A Thousand Plateaus*, trans. Brian Massumi (London: Athlone, 1987), 64.

17. Arthur Schopenhauer, *On the Will in Nature*, trans. Karl Hillebrand (New York: Amazon CreateSpace Independent Publishing Platform, 2010).

18. Friedrich Nietzsche, *The Will to Power in Science, Nature, Society and Art*, trans. Phillip Kaufman (New York: Vintage, 1973).

19. Francis Ponge, "Flora and Fauna," in *The Voice of Things*, trans. Beth Archer (New York: McGraw-Hill, 1971).

20. Mikhail Bakhtin, "Response to a Question from Novy Mir," in *Speech Genres and Other Late Essays*, trans. Vern W. McGee (Austin: University of Texas Press, 1986). We can, for instance, see the Cambrian explosion as something analogous to what Bakhtin calls "Great Times" in this text, generating an exuberance that is reworked and revisited by prior generations.

21. Mikhail Bakhtin, "Author and Hero in Aesthetic Activity," in *Art and Answerability: Early Philosophical Essays*, ed. Michael Holquist and Vadim Liapunov, trans. Vadim Liapunov and Kenneth Brostrom (1924; repr., Austin: University of Texas Press, 1990).

22. Mikhail Bakhtin, *Towards a Philosophy of the Act* (Austin: University of Texas Press, 1993), 40.

23. Félix Guattari, *Chaosmosis: An Ethico-Aesthetic Paradigm*, trans. Paul Bains and Julian Pefanis (Bloomington: Indiana University Press, 1995), esp. 13–18.

24. Asger Jörn's concept of the arabesque is given in the text "Living Ornament." Asger Jörn, "Levend ornament," *Forum: Maandblad voor architectuur en gebonden*

kunsten, no. 4 (1949): 137–47. Jörn's painting *Tunisien* (1948) can be seen as a lively partner to this text.

25. Stengers, *Cosmopolitics*.

26. The United States Forest Service iTree software suite is developed for this purpose.

27. Jaroslav Hašek, *The Good Soldier Schweik*, trans. Cecil Parrott (London: Penguin, 1974).

28. Nietzsche, *Will to Power*, §544.

29. Alexander Vvedensky, *An Invitation for Me to Think*, ed. Eugene Ostashevsky, additional translations by Matvei Yankelevich (New York: New York Review Books, 2013). See also Nadezhda Tolokonnikova's closing defense statement in the Pussy Riot trial of that year that drew on this aspect of Vvedensky's work.

30. Bakhtin, *Art and Answerability*; see also Mikhail Bakhtin, "The Problem of the Text in Linguistics, Philology, and the Human Sciences: An Experiment in Philosophical Analysis," in *Speech Genres and Other Late Essays* (Austin: University of Texas Press, 1986).

31. Daniil Kharms, "Pushkin and Gogol" (written in the 1930s, never published in his lifetime, like most of his other work except children's poetry), available in the public domain and reproduced on multiple sites online (in Russian); translation ours from http://proza.ru/2014/11/08/1591.

32. On Muriel Wheldale, see M. L. Richmond, "The 'Domestication' of Heredity: The Familial Organization of Geneticists at Cambridge University, 1895–1910," *Journal of the History of Biology*, no. 39 (September 2006): 565–605.

33. Italo Calvino, *If on a Winter's Night a Traveller*, trans. William Weaver (1981; repr., London: Vintage, 2010), 68, 238.

34. Nietzsche, *Will to Power*, §543.

35. Francis Hallé, *In Praise of Plants*, trans. David Lee (Portland: Timber, 2011).

36. Dieter Volkmann, František Baluska, Irene Lichtshieldl, Dominique Driss-Ecole, and Gérald Perbal, "Statolith Motion in Gravity-Perceiving Plant Cells: Does Actomyosin Counteract Gravity?," *FASEB Journal*, no. 13 (1999): 143–47.

37. D'Arcy Wentworth Thompson, *On Growth and Form* (Cambridge: Cambridge University Press, 1945). A history of mathematical biology would note its strong relation to other scales of research aside from morphology, notably Thompson's late nineteenth-century work on the threats to populations of animals, such as seals and sea otters, posed by hunting.

38. Tilting plants, having them grow at a nonorthogonal relation to the surface of a growth medium, can almost be said to be a trope in art: Hans Haacke's *Directed Growth* (1972) has a row of runner beans grown in soil on a glass plate on the gallery floor. The plants grow along lines angling away from vertical. Diller Scofidio + Renfro's *Arbores Laetae* (2008) has a pair of young hornbeams, each planted off-center in their own turntable-mounted bed (flush with ground level) rotating

slowly at a tilted angle. Henrik Håkansson has a bed of shrubs in a large irrigated planter, tilted ninety degrees, in his *Fallen Forest* (2006). Images of these works can be found in the catalog *Radical Nature: Art and Architecture in a Changing Planet 1969–2009* (London: Barbican Art Gallery, 2009).

39. Darwin and Darwin, *The Power of Movement in Plants*.

40. R. Ranjeva, A. Graziana, and C. Mazars, "Plant Graviperception and Gravitropism: A Newcomer's View," *FASEB Journal* 13, no. 9001 (May 1999): 135–41.

41. Brian G. Forde and Michael R. Roberts, "Glutamate Receptor-Like Channels in Plants: An Amino Acid Sensing Role in Plant Defence?," *F1000Prime Reports* 6, no. 37 (2014).

42. G. Perbal, "Gravisensing in Roots," in "Life Sciences: Microgravity Research II," special issue, *Advances in Space Research* 24, no. 6 (1999): 723–29. Here, tension in the actin filaments resulting from interaction with the statoliths would be transmitted to stretch-activated ion channels located in the plasma membrane.

43. F. Baluska, Milada Ciamporova, Otília Gasparíková, and Peter W. Barlow, eds., *Structure and Function of Roots: Proceedings of the Fourth International Symposium on Structure and Function of Roots, June 20–26, 1993, Stará Lesná, Slovakia* (Dordrecht: Springer Netherlands, 2013).

44. Francis Ponge, "Moss," in *Unfinished Ode to Mud*, trans. Beverley Bie Brihac (London: CB Editions, 2008), 23.

45. For an affinity for buddleia and the politics of urban life, see Laura Oldfield Ford, *Savage Messiah* (London: Verso, 2011).

46. A. J. Davy, G. F. Bishop, and C. S. B. Costa, "*Salicornia* L. (*Salicornia pusilla* J. Woods, *S. ramosissima* J. Woods, *S. europaea* L., *S. obscura* P. W. Ball & Tutinin, *S. nitens* P. W. Ball & Tutinn, *S. fragilis* P. W. Ball & Tutin and *S. dolichostachya* Moss)," *Journal of Ecology* 89, no. 4 (August 2001): 681–707.

47. United States Department of Agriculture, "Metal-Scavenging Plants to Cleanse the Soil," *AgResearch Magazine*, November 1995. For more on *Revival Field*, see the artist's website, http://melchin.org.

48. Barbara C. Matilsky, *Fragile Ecologies: Contemporary Artists' Interpretations and Solutions* (New York: Rizzoli, 1992), 111.

49. A further interesting development is in the use of plants as biosensors. See Alexander G. Volkov and Don Rufus A. Ranatunga, "Plants as Environmental Biosensors," *Plant Signaling and Behavior* 1, no. 3 (May/June 2006): 105–15.

50. See, for instance, *Serpentine Dance* (1896), the short film by the Lumière brothers documenting Loie Fuller's dance form.

51. Mark C. Mescher, Jordan Smith, and Consuelo M. De Moraes, "Host Location and Selection by Holoparasitic Plants," in *Plant-Environment Interactions*, ed. František Baluška (Berlin: Springer, 2009).

52. Francisco J. Varela, Evan Thompson, and Eleanor Rosch, *The Embodied Mind: Cognitive Science and Human Experience* (Cambridge, Mass.: MIT Press, 1992).

53. Johann Wolfgang von Goethe, *The Metamorphosis of Plants*, with photographs by Gordon L. Miller (Cambridge, Mass.: MIT Press, 2009).

54. Johann Wolfgang von Goethe, *Schriften Zur Morphologie* (1817; repr., Berlin: Hofenberg, 2016).

55. Francisco J. Varela, *Ethical Know How: Action, Wisdom, and Cognition* (Stanford: Stanford University Press, 1999), 57.

56. Varela, 58.

57. Muriel Wheldale, *The Anthocyanin Pigments of Plants* (Cambridge: Cambridge University Press, 1916).

58. Wheldale, 148.

59. Joris-Karl Huysmans, *Against Nature* (London: Penguin, 2003), 98.

60. Maurice Maeterlinck, *The Intelligence of Flowers* (Albany: State University of New York Press, 2007). Maeterlinck gives the lizard orchid the botanical name *Loroglossum hircinum,* and it is also called *Himantoglossum hircinum.*

61. Maeterlinck, 38–39.

62. D. Mollison, Roy Malcolm Anderson, Maurice Stevenson Bartlett, Richard Southwood, Hans Leo Kornberg, and M. H. Williamson, "Modelling Biological Invasions: Chance, Explanation, Prediction [and Discussion]," *Philosophical Transactions of the Royal Society of London B* 314, no. 1167 (1986): 675–93.

63. Carey is also the author of "DISPERSE: A Cellular Automaton for Predicting the Distribution of Species in a Changed Climate," *Global Ecology and Biogeography Letters* 5, no. 4/5 (July–September 1996): 217–26.

64. P. D. Carey, "Changes in the Distribution and Abundance of *Himantoglossum hircinum* (L.) Sprengel (Orchidaceae) over the Last 100 Years," *Watsonia* 22 (1999): 353–64. In this article, which devotes space to reflections on historical and social changes in land use, Carey also notes that the movement of militaries between France and England during World War II may have been an important vector for the migration of seeds.

65. Michel Serres, *Biogea* (Minneapolis: Univocal, 2012), 123.

66. Maeterlinck, *The Intelligence of Flowers,* 23.

67. Francis Ponge, "The (Dried) Fig," in *Unfinished Ode to Mud,* trans. Beverley Bie Brihac (London: CB Editions, 2008), 129–33.

68. John M. Warren, *The Nature of Crops: How We Came to Eat the Plants We Do* (Wallingford: CABI, 2015).

69. Karl Blossfeldt, *Urformen der Kunst* (Berlin: Ernst Wasmuth, 1928); Edward Juler, "The Key to a Hidden World: Photomicrography and Close-Up Nature Photography in Interwar Britain," *History of Photography: An International Quarterly* 36, no. 1 (2012): 87–98. See also Edward Juler, "Life Forms: Henry Moore, Morphology and Biologism in the Interwar Years," in *Henry Moore: Sculptural Process and Public Identity* (London: Tate, 2015).

HOME

1. Anatoly Efros, dir., *On Thursday, and Never Again* (Moscow: Mosfilm, 1977).

2. Pierre-André Boutang and Michel Pamart, dirs., Claire Parnet and Gilles Deleuze, writers, *L'Abécédaire de Gilles Deleuze* (Paris: La Femis Sodaperaga Productions, 1988–89), "A for Animal."

3. V. V. Bibikhin, *Les (hyle): Problema materii, istorija poniatija, zhivaja materija v antichnoj i sovremennoj filosofii* [Forest (hyle): The problem of matter, history of concept, and living matter in ancient and contemporary biology] (St. Petersbourg: Nauka, 2011). Chapters 1, 6, and 8 are published in English as "The Wood(s) on the Problem of Living Matter in Ancient and Contemporary Biology," trans. Michael Marder, *Stasis*, no. 1 (July 4, 2015), http://www.stasisjournal.net. We find Bibikhin's figuration of the forest, which could be read as a kind of cosmic materialism, very suggestive; it has the benefit of relating questions of matter directly to the way in which ideas of matter have been discussed and developed historically, while allowing for a generous proliferation of sites for understanding matter. His project is, however, broader than this question and we are not able to associate with all of it here.

4. Monica Galiano, John C. Ryan, and Patricia Vieira, eds., *The Language of Plants: Science, Philosophy, Literature* (Minneapolis: University of Minnesota Press, 2017).

5. Bibikhin, *Forest (hyle)*, 14, 27, translation adapted by us from *Stasis*, no. 1 (2015), chap. 1.

6. It is impossible to ascertain who actually said this. It likely did not originate from Nabokov; some authors attribute this phrase to the writer Vladimir Korolenko.

7. Home can be expressed in coupled terms. One such pair is bilingualism and English as a Second, Foreign, or, in the British context, Additional Language (EAL). Bilingualism is currently glorious: it is claimed to develop brain plasticity and inventiveness as well as tolerance. The governmental term EAL is less so; it leaks trouble: immigration, asylum, unemployment, poverty, and, generally, lack. The terms do not conjugate: EAL does not seem to mean bilingualism, especially since, for example, as part of the annual school league tables, it is expressed as a percentage, and used to inform parental choice of school. What chimeras, and ideas of homeliness, spring into action upon seeing these figures: 9 percent EAL or 57 percent EAL? Here, the percentage of EAL should presuppose the same dual-language condition, but it does away with the glory of bilingualism: it comes to mean less of each language, without a home language, lacking in both.

8. V. V. Bibikhin, *Sobstvennost'. Filosofija svoego* [Property (ownership): The philosophy of one's own] (St. Petersburg: Nauka, 2012) (a lecture course initially read in 1993-94 and reworked in 1995), 99, 133, translation ours throughout this section.

9. Bibikhin, 133.

10. Bibikhin, 133–34.
11. Bibikhin, 134, 297.
12. Bibikhin, 151, emphasis added.
13. Bibikhin, 113.
14. Bibikhin, 284–85.
15. Bibikhin, 284–85, 138–41.
16. Bibikhin, 344.
17. Bibikhin, 351.
18. Martin Heidegger, *Being and Time* (New York: New York University Press, 1996), 256–66.
19. Bertolt Brecht, "On Hegelian Dialectics," 1940–41, http://www.autodidactproject.org.
20. Rosi Braidotti, "The Contested Posthumanities," in *Conflicting Humanities*, ed. Rosi Braidotti and Paul Gilroy (London: Bloomsbury, 2016), 15–20.
21. James Tully, *An Approach to Political Philosophy: Locke in Contexts* (Cambridge: Cambridge University Press, 1993); Barbara Arneil, *John Locke and America: The Defence of English Colonialism* (Oxford: Oxford University Press, 1994).
22. Étienne Balibar, *Identity and Difference: John Locke and the Invention of Consciousness* (London: Verso, 2013).
23. Brenna Bhandar, "Title by Registration: Instituting Modern Property Law and Creating Racial Value in the Settler Colony," *Journal of Law and Society* 42, no. 2 (June 2015): 253–82.
24. Alexei Yurchak, *Everything Was Forever, Until It Was No More: The Last Soviet Generation* (Princeton, N.J.: Princeton University Press, 2006).
25. An insightful overview of Deleuze's relation to the work of Heidegger can be found in Constantin Boundas, "Martin Heidegger," in *Deleuze's Philosophical Lineage*, ed. Graham Jones and Jon Roffe (Edinburgh: Edinburgh University Press, 2009).
26. Gilles Deleuze and Félix Guattari, *A Thousand Plateaus*, trans. Brian Massumi (London: Athlone, 1987), 343.
27. Deleuze and Guattari, 354.
28. Deleuze and Guattari, 372–73.
29. Deleuze and Guattari, 378.
30. Gaston Bachelard, *The Poetics of Space* (Boston: Beacon, 1994), 5.
31. See Deleuze and Guattari, *A Thousand Plateaus*, 377.
32. Andrei Tarkovsky, dir., *Solaris* (Moscow: Mosfilm, 1972).
33. Deleuze and Guattari, *A Thousand Plateaus*, 378.
34. Deleuze and Guattari, 359.
35. Lodge Hill, Kent, owned by the Ministry of Defence, is the largest UK breeding site for nightingales and is the subject of a plan to build five thousand houses.
36. Deleuze and Guattari, *A Thousand Plateaus*, 375.

37. Rosi Braidotti, *Nomadic Subjects: Embodiment and Sexual Difference in Contemporary Feminist Theory* (New York: Columbia University Press, 2011).

38. Homi K. Bhabha, *The Location of Culture* (New York: Routledge, 2004), 15.

39. For a discussion of some of these tensions discussed in relation to the political life span of the Organisation of Women of Asian and African Descent, an activist group of the late seventies and early eighties, see Brixton Black Women's Group, "Black Women Organising," April 17, 2017, http://libcom.org/library/black-women-organising-brixton-black-womens-group.

40. Ágota Kristóf, *The Illiterate* (London: B Editions, 2014), 34.

41. Ágota Kristóf, *Yesterday* (London: Vintage, 1997).

42. Mark Fisher, "Alien Traces: Stanley Kubrick, Andrei Tarkovsky, Christopher Nolan," in *The Weird and the Eerie* (London: Repeater Books, 2016).

43. W. G. Sebald, *The Emigrants* (London: Vintage, 2002), 56.

44. Félix Guattari, *Three Ecologies* (London: Bloomsbury, 2005), 37, 42–43.

45. Iain Chambers, *Migrancy, Culture and Identity* (New York: Routledge, 1994); Avtar Brah, *Cartographies of Diaspora: Contesting Identities* (New York: Routledge, 1996).

46. Hannah Arendt, *The Human Condition* (Chicago: University of Chicago Press, 1958), 30.

47. Bachelard, *The Poetics of Space*, 6.

48. Bachelard, 16.

49. Bachelard, 7.

50. Bhabha, *The Location of Culture*, 18–19.

51. Adrienne Rich, *Of Woman Born: Motherhood as Experience and Institution* (New York: Norton, 1995); Jacqueline Rose, *Mothers: An Essay on Love and Cruelty* (London: Faber & Faber, 2018).

52. Isabelle Stengers, "Matter of Cosmopolitics: On the Provocations of Gaia," in *Architecture in the Anthropocene: Encounters among Design, Deep Time, Science and Philosophy*, ed. Etienne Turpin (London: Open Humanities Press, 2013), 178.

53. Andrei Tarkovsky, dir., *Mirror* (Moscow: Mosfilm, 1974).

54. Alexander Sokurov, dir., *Mother and Son* (Russia and Germany: Roskomkino, North Foundation, Zero Film, Lenfilm, 1997).

55. Alexander Etkind, *Internal Colonization: Russian Imperial Experience* (Cambridge, Mass.: Polity, 2011).

56. John Locke, *Second Treatise of Government* (London, 1690; Project Gutenberg, 2010), http://www.gutenberg.org.

57. Eyal Weizman (text) and Fazal Sheikh (photographs), *The Conflict Shoreline: Colonization as Climate Change in the Negev Desert* (Göttingen: Steidl, 2015).

58. Weizman and Sheikh, 52.

59. Weizman and Sheikh, 10–11.

60. Weizman and Sheikh, 24.

61. Weizman and Sheikh, 48.
62. Weizman and Sheikh, 12.
63. See Maria Thereza Alves, *Decolonizing Brazil,* ebook available on the artist's website, http://www.mariatherezaalves.org; see also the project website, http://www.descolonizandobrasil.com.br/#.
64. James C. Scott, *The Art of Not Being Governed: An Anarchist History of Upland Southeast Asia* (New Haven, Conn.: Yale University Press, 2009); James Koehnline, *Gone to Croatan: Origins of North American Dropout Culture* (New York: Autonomedia, 1994).
65. Cary Wolfe, "The Poetics of Extinction" (lecture, Goldsmiths University of London, June 13, 2015).
66. Peter Linebaugh, *The Magna Carta Manifesto: Liberties and Commons for All* (Berkeley: University of California Press, 2008).
67. Linebaugh, 36.
68. Etkind, *Internal Colonization,* 81.
69. Etkind, 78–79.
70. Etkind, 84, 83.
71. Etkind, 81.
72. Oleg Vilkov, "Pushnoi Promysel v Sibiri," *Nauka v Sibiri,* no. 45 (November 19, 1999), http://www.nsc.ru/HBC/article.phtml?nid=165&id=19.
73. Etkind, *Internal Colonization,* 87.
74. A. Shchapov, *Sochinenija* [Written works], vols. 1–3 (St. Petersbourg: Izdanie M. V. Pirozhkova, 1906).
75. Etkind, *Internal Colonization,* 66. See also Shchapov, *Sochineniya,* esp. "Istoriko-geograficheskoe raspredelenie russkogo narodonaselenija" (first published in 1864).
76. Shosuke Yoshida, Kazumi Hiraga, Toshihiko Takehana, Ikuo Taniguchi, Hironao Yamaji, Yasuhito Maeda, Kiyotsuna Toyohara, Kenji Miyamoto, Yoshiharu Kimura, and Kohei Oda, "A Bacterium That Degrades and Assimilates Poly(ethylene terephthalate)," *Science* 351, no. 6278 (March 11, 2016): 1196–99.

Index

Abraham, 59, 74
Abstract Expressionism, 61
aesthetic(s), xi–xii, xvi–xix, xxii–xxvi,
　11, 14, 28, 31, 37–38, 43, 45, 77,
　93–94, 98–99, 138–39, 143, 157–58;
　activity of plants and vegetal, xix,
　93, 101–2, 108; of becoming and
　of uncertainty, 112; of chance, xvii,
　79; condition and mode, 73–74; of
　domination, 82; ecological, xi–xiii,
　xviii, xxviii, 151; entity, 101; event,
　xxvi, 14, 32; experience, xxii, 29,
　31, 38
affect, 30–31, 61, 96
Aion, 80, 90
Alexievich, Svetlana, 51
Ammophila. See marram grass
Amoore, Louise, 64
anguish, xv, xxii–xxiii, 26–49, 72,
　78, 98, 102, 123–24, 136–37; of
　motherhood, 140–41; Sartre's
　conception of, 59
animal, xxi, 28, 42–43, 45–46, 78, 91,
　108, 121–23, 130, 150–52, 159; and
　adaptive suicide, 40
Antirrhinum majus. See snapdragons
aphid, 40–41

Appadurai, Arjun, 89
Aquinas, Thomas, 44
arabesque, xix, 101, 111–13, 118–19
Arcimboldo, Giuseppe, 113
Arendt, Hannah, xx, 137–39
arms race, 58, 62, 64
Arp, Jean, 113
art, xviii–xix, xxiv–xxv, 43, 55; of living
　(see also *metis*), xxiii, 83, 92, 99; of
　statecraft, 73; of war, 14, 76
Art Nouveau, 101, 112
Atkins, Ed, 62
atomic: bombs and weapons, 56, 58;
　energy, 122; particles, 56, 151. *See
　also* nuclear
automata, xvii, 60, 66–68, 70, 72, 74
Averintsev, Sergei, 48

Bachelard, Gaston, xx, 138–39, 143
bacteria, 5, 10, 20, 152–53; Ideonella
　sakaiensis 201-F6, 153
Bakhtin, Mikhail, xii, xix, xxiv–xxvii,
　99–102, 105, 158; *Aesthetics and
　Theory of the Novel*, xxv; *Art and
　Answerability*, xxv; *The Problem of
　the Text*, xvii
Balestrini, Nanni, 73–74

Balibar, Étienne, 128
Baluška, Frantisek, xviii
Barthélémy, Jean-Hugues, 4
Bateson, Gregory, xvi, 28, 60, 94, 101
Bateson, William, 106
Baudrillard, Jean, xvii, 77, 81–83, 87, 92
beautiful soul, figure of, 2, 14
Beckett, Samuel, 51
becoming, xiii–xv, xxv, xxvii, 1–3, 5–7, 31, 37–38, 44, 84–85, 88, 97, 112, 144, 156, 158; away, 136; of devastation, xxii, 1–3, 5–7, 11, 13, 15–18, 22–23; fatalistic force of, 106; pure, 80–81; in Sartre, 59
Bedouin, 146–47
Beer, Stafford, 94
being, xx, xxiii–xxiv, 2, 29, 83, 130, 151; ethico-aesthetic, 100–101; Hegelian, 126–27, 129; Heideggerian, 55, 127, 129
Berardi, Franco "Bifo," 81, 87
Bergson, Henri, 2, 80
Berlant, Lauren, 20–21, 27
Bhabha, Homi K., xx, 134, 139
Bhandar, Brenna, 128
Bhopal gas leak, 11
Bibikhin, Vladimir, xx, 122–23, 125–29, 132, 137, 143–44, 151–52; *Forest (hyle)*, 122–24; *Property (Ownership)*, 125
biology, 4, 18, 93, 97, 122
black soil, 145–46
Blossfeldt, Karl, 118
Böll, Heinrich, 43
Boshier, Derek, 61
botany, xix, 94, 102, 108
Braidotti, Rosi, xii, 128, 132
Braque, Georges, 157
Brassier, Ray, 4

British Petroleum (BP), 7, 40
Burroughs, William, 87

Calvino, Italo, 107
camphor tree (*Dryobalancops aromatica*), 96
capitalism, xxiii, 26, 54, 147; cognitive, 46, 49; industrial, 149
Carey, P. D., 118
Carson, Rachel, 15
Carthage, 14–15, 17
catallaxy, 55, 68–69, 71
cellular suicide, 4; adaptive, 40
chance, xvii–xviii, xxiv, 47, 73–92, 99, 130, 151, 159; structurations of, xxii, 76, 79, 87
Chaney, Rufus, 110
chaosmosis, 77, 130
Charter of the Forest, 148
Chekhov, Anton, 48
Chernobyl, 4–6, 28, 42, 88
Chin, Mel, xix, 110–11
choice, xvi–xvii, 51–52, 54–55, 59–60, 74; impossible and wrong, xvi, 54, 58–59, 74; structures of and mechanisms of, 55, 57–58
Chronos, 80, 90
climate: as a project for colonial power, 147
climate damage, xxii–xxiii, 7–8, 13, 16, 43, 48, 53, 54
Codariocalyx motorius. See dancing plant
Cold War, xvi, 18, 51, 54–57, 63–65, 72, 80, 85
colonialism: agricultural, 128, 146; fishing, 150; imperial, 147; plant-based, 145; settler, 128, 146
commons, xiv; of anguish, 46; necro-, 22–23; of pollutants, 10–11, 21, 48; theft of, 148
computation, 62, 67, 70, 90, 94, 157

INDEX

conatus, xii, xv, 20
Connolly, William, 69
cosmos, xii, 12, 48, 81, 133, 138, 144, 157; in Deleuze and Guattari, xvii, 130–31
creeping cedar (*Pinus pumila*), 35
cubism, 157
cultivation, xx, 124, 146–47, 151
Cusk, Rachel, 51
cybernetics, xiii, xxiv, 94, 98; second-order, 67

dancing plant (*Codariocalyx motorius*), 97
Dardot, Pierre, 70
Darwin, Charles, xvii, 81, 95, 107, 109
death: in Bartéhlémy's interpretation of Simondon, 4; in Deleuze, 121; in devastation, 4–6; diseases and, 46–47; in Freud, xiv; in Heidegger, xx, 127, 130; in Kierkegaard, xvii; and Muratova, 26, 41; new forms of, 10
de Cusa, Nicholas, 44
Deepwater Horizon, xiv, 7, 40, 78
Deleuze, Gilles, xii, xv, xvii, 14, 33–34, 37–38, 48, 63, 77, 84, 129; *Abeécédaire*, 121; Deleuzian ontology, 44–46; *Difference and Repetition*, 2; *Logic of Sense*, 79, 80–83; *Nietzsche and Philosophy*, 81; *Pure Immanence*, 33, 37, 44; "Tenth Series of the Ideal Game," 81
Deleuze and Guattari, xx, xxiv, 27, 44, 82, 97, 123, 129–32; "Of the Refrain," 130; *A Thousand Plateaus*, xxiv, 32
Derrida, Jacques, 148
Descartes, René, 2; Cartesian idiot, 48
Des Esseintes, Jean, 115–17
destructive plasticity, xiv, 9, 12–13

deterrence, xvi–xvii, 54–55, 57–58, 60, 64, 71, 72, 98
devastation, xiii–xiv, xxii–xxiii, 1–18, 20, 22–23, 28–31, 35, 39–40, 46, 52, 67, 71, 78, 123–24, 133, 137, 159
dialogic, xxvii, 99–100, 157
difference, xxiv, 2, 12, 37, 99, 129, 133, 139
Dionaea muscipula. *See* Venus flytrap
dispossession, 124, 128, 146–47
dodder plant (*Cuscuta*), 112
Dostoevsky, Fyodor, 48, 51
Dryobalancops aromatic. *See* camphor tree
dwelling: in Heidegger, 127, 130, 139

Eco, Umberto, 73
ecology, xviii, xx–xxi, xxiii–xxiv, 2, 16–17, 19, 22, 90, 98, 122–24, 156; of anguish, 30, 42; chance in, 78; deep, 45; ecological materialism, xii; and economy, 148; in Guattari, 1; informational, xvii; of mind, 28; beyond "nature," 1
economy, 8, 17, 67, 87, 136; of chance, 79; of deterrence, xvi, 54; of happiness and of anguish, 46; modeling, 63; of pain, 141; political, xvii, xxii; socioeconomic, 56; zoological, 150
Efros, Anatoly, 121–22, 142, 144
enactivism, 112
enclosure movement, 149
entities, xiii, xxi, 1, 5, 57, 95, 99, 114, 157–58; aesthetic, 101; anguished, 41, 43, 47; murmuring, 63; non-, 71; in Whitehead, 31
environment: and experience of anguish, 38; living, 123; and organism, xix, xxvii; and plant, 114; and plant intelligence, 94–95, 97

INDEX

Eriugena, Johannes Scotus, 44
ethico-aesthetics, xii, xv, xx–xxv, xxviii, 43, 83, 158; approach, xxvii, 91, 99, 101–2, 110; and chance, 76–77, 79, 83, 85, 88; description of, xvii, 3, 6, 26–27, 48, 76, 77; dimensions, 77, 124; ecological, 21, 98; in Guattari and Bakhtin, xii, xxiv–xxvi, 99–100; and home, 130, 133; of plants, 94, 98, 101–3, 108, 118–19; sensibilities, xv, 28, 31, 48
ethics, xxiii–xxiv, 17, 42, 158
Etkind, Alexander, 145, 149–50
eucalyptus, 147
European Central Bank, 64
exclusion zone, 4, 6
existentialism, 27, 169
extinction, xxi, 1, 5–6, 18, 53

fatalism, xix, 27, 79, 103–5, 107–8, 110–11, 114
fate, xvii–xviii, xxii, 39, 74, 76–79, 82, 84, 86–89, 92, 104–5, 124, 141
feelings, 38, 47, 59, 103; negative, 27; in Whitehead, 30–31
feminism, 128, 134, 137–38, 141
Fisher, Mark, 135
Flaubert, Gustav, 59, 60, 73
flower, 32, 93, 113–19, 143, 159–60
Forbes, Scott, 40
forest, xi, xii, xix–xxi, xxiii, 33, 76, 90, 121–24, 129, 139, 142–53; bush, 148; and desert, 146, 147; jungle, 147–48, 152; rainforest, 147; Russian, history of, xx, 149; Siberian, 149–50; wilderness, highlands, and marshlands, 148
Forsythe, William, 111
Foucault, Michel, 38, 63, 70
Freud, Sigmund, xiv
Fuller, Loie, 112

fungi, 42, 93, 151; microrhizoids, 96
fur trade, 124, 149–50

Galton board, xviii, 84–86, 91
game theory, 54, 56–57, 60–61, 63, 65, 71, 73
gas: flaring and leak, 8, 11
Gasparov, Mikhail, 42–43
ghost, 63–64, 67; ghost town, 71–72
glory, xix, 27, 103, 111, 113–18
god and gods, 45, 80, 83–84, 88; Abraham's, 59; Christian, 37; Dionysus, glorious, 31; goddess, xxiii; Leibniz's, 159; Spinoza's, 131
Goedel, Kurt, 62
Goethe, Johann Wolfgang von, 113–14
Gogol, 105–6
Goncharov, Ivan, 51
governance, 47, 57; colonial, 128; neoliberal, 70; preemptive, 64; probabilistic, 74
gravity, xviii, 97, 108–9; *gravisensing*, 108
Growth and Form (exhibition), 113
Guattari, Félix, xii, 1, 20, 27, 81, 100, 136, 158; *Cartographies Schizoanalytiques*, xxv–xxvi; *Chaosmosis*, xxiv–xxvi
Gulag, xv; camp, 35

Hall, Matthew, 93
Hallé, Francis, xix, 108
Hamilton, Richard, 113; *Heteromorphism*, 113; *Microcosmos: Plant Cycle*, 113; *Self Portrait*, 113
Haraway, Donna, 6
Hašek, Jaroslav, 105
Hayek, Friedrich A., xvi, 67–70; *The Constitution of Liberty*, 68; *Law, Legislation and Liberty*, 68–69
Hegel, Georg Wilhelm Friedrich, xx, 2, 9, 34, 52, 125–29, 131–32, 137

INDEX

Heidegger, Martin, xx, 55, 127, 129–30, 132, 139
high-fructose corn syrup (HFCS), 1, 21
Himantoglossum hircinum. *See* lizard orchid
Hippocrates, 118
home, xviii–xxi, xxiii, xxvi, 79, 121–53; homeland, 54, 129–31, 135–36, 152
human rights, 42
Huysmans, Joris-Karl, 115

Ibsen, Henrik, 48
idealism: German, 128, 137; Hegelian, 127
Ideonella sakaiensis 201-F6. *See under* bacteria
immanence, xxiv, 38; ontology of, 33, 44–46; philosophy of, xv
indifference, 11, 16, 65, 98, 104
individuation, xv, xxvii, 4, 30, 37–40, 71, 105, 115, 157–58
infrastructure of feeling, xvi, 53–54, 65–66, 71, 73
intelligence, xviii, 94–95, 97–99, 117, 137, 159–60
International Atomic Energy Authority, 58
International Monetary Fund, 64
Iraq, invasion of, 58
irresolvability, xv–xvii, xxii–xxiii, 12, 17–18, 36, 40, 52–65, 69–74, 76, 79, 98, 124, 133, 137

Jarry, Alfred, 14
Jelinek, Elfriede, 51
Jewish National Fund, 147
JODI, 62

Kairos, 74, 99
Kane, Sarah, xvii, 74
Kant, Immanuel, xxv, 9, 32

Keynesianism, 77
Kharms, Daniil, 105
Kierkegaard, Søren, xvi, xvii, 58–59, 74
king's holly (*Lomatia Tasmanica*), 108
Klee, Paul, 113
Klein, Melanie, xxvi
Krasznahorkai, László, 51
Kristeva, Julia, 27
Kristóf, Ágota, xx, 134–36

labor, 128, 149–50; agrarian, xx, 146–47; vegetal, 103
Lacan, Jacques, xxvi
Lamarck, Jean-Baptiste, 118
law, 18, 134; and Baudrillard, 82, 87; and possibility, 9, 92; property, 126; and violence, xxi
Lazzarato, Maurizio, xxvii
Leandre, Joan, 62
Leibniz, Gottfried Willhelm, 159
life, vegetal, xxiii, 93, 102
Linebaugh, Peter, 148–49
Linnaean taxonomy, 114
literature, xxv, xxvii, 28, 48, 51, 55, 73, 93, 99–100, 106; being at home in, 123–25, 158
lizard orchid (*Himantoglossum hircinum*), 117
Locke, John, 128, 146
Lomatia Tasmanica. *See* king's holly
Lowry, Malcolm, 51
luck, xvii–xviii, xxii, 32, 78–79, 82, 84, 87, 89–92, 107, 124; bad, xviii, 36, 42, 75–76, 79, 83, 91; good, 91; worse, 91–92

Maeterlink, Maurice, 117–18
Magna Carta, 148
Malabou, Catherine, xiv, 9, 12, 18, 29
Mallarmé, Stéphane, 81
Marder, Michael, 93

market, 21, 66–71; in Hayek, 55, 68–69
Markson, David, 51
Markvoort, Eva, 47
marram grass (*Ammophila*), 103
marsh samphire (*Salicornia europaea*), 110
Massumi, Brian, 64
Matrosov, Alexander, 39
McClintock, Barbara, xix, 106, 108
Medea, xvii, 74
Mendel, Gregor, 106, 108
metaphysics, 157–58
metis, xxiii, 14, 17, 79, 83, 92, 99, 152
migrant, xx, 113, 123, 125, 135–39, 141–42
migration, 131, 133, 135–36, 152
military, 85; bases and technologies, 56–57; becoming, 1; as gaming field, 16; systems and bluff, 73; uses of ecological problems, xiv
Mimosa pudica. See sensitive mimosa
mind, theory of, 73
Mirowski, Philip, 63, 67
Monte Testaccio, 156
Morgenstern, Oskar, 56. See also von Neumann, John
mother, 39; and motherhood, 139–42
Muratova, Kira, xv, 11, 26, 29, 33, 36, 41–45; *The Asthenic Syndrome*, 26; *Three Stories*, xv, 26, 41, 45, 47
Muscovite state and formation, 144, 149–50

Nabokov, Vladimir, xv, 28, 125
natal, xx, 129–32, 136, 142, 148
nation, xx, 89–90, 121, 125–26
nature, xiv–xvii, xx, 15, 19, 34, 115–17, 144, 147; natural, 1, 5–6, 40, 117, 138, 147, 149, 157; postnatural, 117
Nealon, Jeffery, 93
Negev Desert, 146–47

Ngai, Sianne, 27
Nietzsche, Friedrich, xii, xv, xvii, 14, 33–37, 39, 45, 48, 77, 80–84, 97, 104–5, 107
Niger Delta, xiv
Nixon, Richard, 21, 65
nonhuman, 28, 40–41, 45, 48, 93
Noys, Benjamin, 2
nuclear: annihilation, 60; bomb, 61; condition, 55; deterrence, 58, 98; infrastructure, 56–58, 88; warfare, xxii; weapons, 17–18, 65. See also atomic

oikos, 148
oil, 7–8, 11, 122, 152; spill, xiv, 1, 7–8, 40, 89
O'Keeffe, Georgia, 119
ontology: Deleuzian, 45; Deleuzoguattarian, xxiv; as dice throw, 81; Heideggerian, 127; of immanence, 33, 44; ontological load, xvii, 76–77, 79, 84, 90, 156; Shalamov's, 35
Ostrom, Elinor, 10
ownership, 125–26, 128–29, 143–44, 146, 148, 150

parent. See mother
partial object, xxvi, 101
Paxton, Joseph, 147
Pennycress (*Thlaspi*), 110
Picasso, Pablo, 157
Pig's Eye Landfill Site, 110
Pinus pumila. See creeping cedar
plant, xviii, xix, xxi–xxiii, 27, 38, 43, 78–79, 89, 93–99, 101–19, 122, 144–48, 150, 151, 159, 160; Aristotelian and Linnaean view, 108; behavior, 96; communication, 95–96; cunning, 98–99, 105, 109,

INDEX

112; imperialism, 147; intelligence, xviii, 94–95, 97–99, 117; plant-based colonialism, 145; politics of planting, 124, 145–46; roots, xi; sensing, 96, 98; temporality, 108; as a weapon, 147; will, 97
plastic, 10–12, 22, 48; commons of, xiv, 10; micro-, 10, 22; plasticizer, 22; polyethylene terephthalate (PET), 153; waste, 153
plasticity, destructive, xvi, 12, 18, 29, 30
Pop Art, xvi, 61, 62
postcolonial, 128, 137–39
preemption, 64
probability, xviii, 65, 76–78, 81, 84, 87; theories of, 77, 87
property, xx, 69, 125–26, 128–29, 142–43, 148, 151
Pushkin, 105–6

radiation, xiii, 3–6, 75–76
Rich, Adrienne, 140
risk, xvii–xviii, xxii, 7, 75–79, 84, 86, 88, 90, 92, 94
Romanticism, 128, 132
root (of plant), xviii, 96, 103, 107–13, 159
Rosch, Eleanor, 112
Rose, Jacqueline, 104
rose of Jericho (*Selaginella lepidophylla*), 104
Rosset, Clément, 34, 84, 92
Russell, Bertrand, 58

Salicornia europaea. *See* marsh samphire
Sartre, Jean-Paul, xvi, 27, 55, 58–60, 62–63
Schelling, Thomas, 57
schizophrenia, xvi, 60
Schopenhauer, Arthur, 34, 37, 51, 55, 97

science, xiii, xxv–xxvi, 6, 15–16, 82, 114; cognitive, 61; ecological, 13; human, xxv; neuro-, 61; plant, xiii
Scotus, John Duns, 44
Sebald, W. G., xx, 135–36
Selaginella lepidophylla. *See* rose of Jericho
self-organization, 54, 67–72
sensitive mimosa (*Mimosa pudica*), 97
serfdom (in Russia), 124, 145–46
Serres, Michel, 14, 118
Shalamov, Varlam, xv, 33, 35–36
Shannon, Claude, 66
Shaviro, Steven, 32
Shaw, Bernard, 48
Shchapov, Afanasiy, 150
Sheikh, Fazal, 146
Simondon, Gilbert, 4
snapdragons (*Antirrhinum majus*), 114
Socrates, 34, 37; pre-Socratic, 122
Sokurov, Alexander, 144
Solzhenitsyn, Alexander, 34–35
Sontag, Susan, 27
species, xiii–xiv, xvi, xviii–xix, 10, 15, 19–20, 22, 29, 36, 76, 78, 89, 94, 96–97, 101, 104, 106, 109, 112, 114–15, 118, 123, 125, 141, 149–50, 158–60; extinction of, xxi, 2, 6; inter- and cross-, 21, 48; multi-, 93
Spinoza, Baruch, xii–xiii, xv, 23, 29, 31, 44, 131
Stasi, 54
Stengers, Isabelle, 102, 141
Steyerl, Hito, 62
Stiegler, Bernard, 14
structures of feeling, 54
Strugatsky, Boris and Arkady, 11
subject, xiii, xv, 11–12, 14, 16, 26, 29–32, 38–39, 44, 52–53, 57, 59–63, 141, 143, 157; agency of, 28; in

Braidotti, 132; and chance, 87–88; formation of, 70–74; in Guattari, 136; intersubjectival, xxvi; and object, 44; in Sartre and Flaubert, 59–60, 62; self-possessing, 128; sovereign subjecthood, 26–27; subjectival life and infrastructure, 55–57, 73, 102; subjectivity and subjectivation; xi, xvi, xxii, 20, 23, 54, 57, 132–33; in Whitehead, 30

systems theory, xiii, 94, 98

Tanning, Dorothea, 119
Tarkovsky, Andrei, xix, 122, 133, 143–45; *Mirror*, 121, 144; *Nostalgia*, 121; *The Sacrifice*, 121; *Solaris*, 121, 131, 135, 144; *Stalker*, 11, 121
technology, computational; 47, 55, 62, 67, 70, 90, 94, 113, 157; and Heidegger, 55
territory, 89, 136, 142; of animal in Deleuze and Guattari, 130–32; control over, 17, 74, 150; of death in Deleuze and Parnet, 121–23
Thacker, Eugene, xv, 7, 44–46
Thatcher, Margaret, 68
Third Reich, 54
Thlaspi. *See* pennycress
Thompson, D'Arcy Wentworth, 108, 113
Thompson, Evan, 112
toska, xv, 28–29, 33, 38, 43, 48
toxin, 10, 110
transitional object, xxvi. *See also* partial object
Trecartin, Ryan, 62
Trewavas, Anthony, xviii, 95
Tsing, Anna, 9

unhomely, 134–36, 139, 142

value, 26, 33–35, 48, 148; in Hayek, 68, 70–71
van Mises, Ludwig, 68
Varela, Francisco, 112, 114
Venus flytrap (*Dionaea muscipula*), 97
virtual/virtuality, xiii, 1–4, 6, 9, 12, 18, 28, 30–31, 35, 38–39, 45–46, 73, 81
vitalism, devastating, 3, 33
vitalist: ontology, 33, 44; philosophy, 27. *See also* immanence
void, 6, 8–9, 46, 53, 130
von Clausewitz, Carl, 17
von Neumann, John, 56, 57, 66–67, 70, 72
von Trier, Lars, xiv, 6
Vvedensky, Alexander, 105

waste, xiv, 1; commons of plastic, 10–11; plastic, 153; toxic, 13, 48
Weizman, Eyal, 18, 146–47
Wheldale, Muriel, 106, 108, 114
Whitehead, Alfred North, 2, 30–32
will: Hegelian, xx, 126, 131; in Schopenhauer, Nietzsche, and Ponge, 97
Williams, Raymond, 54
Winnicott, D. W., xxvi
witness, xiv, xxii, 12–16, 18, 29, 35–36, 46–47, 76
Wolf, Christa, xv, xvii, 43, 51–52, 54–56, 58, 61; *Medea*, 74; *Parting from Phantoms*, 148
Wolfe, Cary, 148
wood, 122, 143, 148. *See also* forest

Yurchak, Alexei, 128

Zarathustra, 80–81, 91
zinc, 108–10

(continued from page ii)

41 *Matters of Care: Speculative Ethics in More Than Human Worlds*
   Maria Puig de la Bellacasa

40 *Of Sheep, Oranges, and Yeast: A Multispecies Impression*
   Julian Yates

39 *Fuel: A Speculative Dictionary*
   Karen Pinkus

38 *What Would Animals Say If We Asked the Right Questions?*
   Vinciane Despret

37 *Manifestly Haraway*
   Donna J. Haraway

36 *Neofinalism*
   Raymond Ruyer

35 *Inanimation: Theories of Inorganic Life*
   David Wills

34 *All Thoughts Are Equal: Laruelle and Nonhuman Philosophy*
   John Ó Maoilearca

33 *Necromedia*
   Marcel O'Gorman

32 *The Intellective Space: Thinking beyond Cognition*
   Laurent Dubreuil

31 *Laruelle: Against the Digital*
   Alexander R. Galloway

30 *The Universe of Things: On Speculative Realism*
   Steven Shaviro

29 *Neocybernetics and Narrative*
   Bruce Clarke

28 *Cinders*
   Jacques Derrida

27 *Hyperobjects: Philosophy and Ecology after the End of the World*
   Timothy Morton

26 *Humanesis: Sound and Technological Posthumanism*
David Cecchetto

25 *Artist Animal*
Steve Baker

24 *Without Offending Humans: A Critique of Animal Rights*
Élisabeth de Fontenay

23 *Vampyroteuthis Infernalis: A Treatise, with a Report by the Institut Scientifique de Recherche Paranaturaliste*
Vilém Flusser and Louis Bec

22 *Body Drift: Butler, Hayles, Haraway*
Arthur Kroker

21 *HumAnimal: Race, Law, Language*
Kalpana Rahita Seshadri

20 *Alien Phenomenology, or What It's Like to Be a Thing*
Ian Bogost

19 *CIFERAE: A Bestiary in Five Fingers*
Tom Tyler

18 *Improper Life: Technology and Biopolitics from Heidegger to Agamben*
Timothy C. Campbell

17 *Surface Encounters: Thinking with Animals and Art*
Ron Broglio

16 *Against Ecological Sovereignty: Ethics, Biopolitics, and Saving the Natural World*
Mick Smith

15 *Animal Stories: Narrating across Species Lines*
Susan McHugh

14 *Human Error: Species-Being and Media Machines*
Dominic Pettman

13 *Junkware*
Thierry Bardini

12 *A Foray into the Worlds of Animals and Humans,* with *A Theory of Meaning*
Jakob von Uexküll

11 *Insect Media: An Archaeology of Animals and Technology*
   Jussi Parikka

10 *Cosmopolitics II*
   Isabelle Stengers

9 *Cosmopolitics I*
   Isabelle Stengers

8 *What Is Posthumanism?*
   Cary Wolfe

7 *Political Affect: Connecting the Social and the Somatic*
   John Protevi

6 *Animal Capital: Rendering Life in Biopolitical Times*
   Nicole Shukin

5 *Dorsality: Thinking Back through Technology and Politics*
   David Wills

4 *Bíos: Biopolitics and Philosophy*
   Roberto Esposito

3 *When Species Meet*
   Donna J. Haraway

2 *The Poetics of DNA*
   Judith Roof

1 *The Parasite*
   Michel Serres

MATTHEW FULLER is professor of cultural studies at Goldsmiths, University of London.

OLGA GORIUNOVA is reader at Royal Holloway, University of London.

CPSIA information can be obtained
at www.ICGtesting.com
Printed in the USA
BVHW042243261119
564894BV00012B/347/P